Digital Political Participation, Social Networks and Big Data

José Manuel Robles-Morales ·
Ana María Córdoba-Hernández

Digital Political Participation, Social Networks and Big Data

Disintermediation in the Era of Web 2.0

José Manuel Robles-Morales
Departamento de Sociología III
Universidad Complutense de Madrid
Madrid, Spain

Ana María Córdoba-Hernández
Departamento de Comunicación
Pública
Universidad de La Sabana
Chía, Colombia

ISBN 978-3-030-27756-7   ISBN 978-3-030-27757-4   (eBook)
https://doi.org/10.1007/978-3-030-27757-4

Cover credit: Ikon Images/Alamy Stock Photo

This Palgrave Macmillan imprint is published by the registered company Springer Nature Switzerland AG
The registered company address is: Gewerbestrasse 11, 6330 Cham, Switzerland

For Carlos, José, María Cristina and Carmen

Contents

LIST OF FIGURES

LIST OF TABLES

CHAPTER 1

Introduction

It is not easy to write a book like this. Among the reasons we could give would be how easy it is to make the same mistake as Fabrizio del Dongo. At one point, this character from Stendhal's novel "The Charterhouse of Parma" finds himself in the middle of the battle of Waterloo. Amid the whirl of activity all around him, he asks insistently where the battle is taking place. From his point of view, he could only see soldiers coming and going, a few skirmishes, lots of noise, but not the battle itself. His subjective position meant he could not grasp the full meaning of the tremendous movement of people and processes taking place around him.

When researching communication and the Internet, we think about how political parties and the media are adapting to the new digital culture and language, but also about how citizens adopt digital technologies in their lives, adapting and transforming them until they can be used for a purpose they were not designed for: political messages. In discussing something as alive, changing and elusive as digital communications, something that generates innovation every day (new types of messages, more specific applications, new devices, etc.), we feel a little like Stendhal's character must have felt. A feeling that leads us to think that all we can capture with our tools is no more than an epiphenomenon of something much larger.

There is a persistent feeling that there is more happening out there, that a new application is about to change the way we understand communication and that it will force us to rethink everything; to go back and start

© The Author(s) 2019
J. M. Robles-Morales and A. M. Córdoba-Hernández,
Digital Political Participation, Social Networks and Big Data,
https://doi.org/10.1007/978-3-030-27757-4_1

again. This sensation, often sustained by adspeak or tech gurus, is based on the unshakeable conviction that our society is changing, and we must therefore be open to all types of innovation. One of these innovations is called *disintermediation*. What is this disintermediation?

Disintermediation is a process that may be reversing a key aspect of our cultural, social and economic models. The rise of industrial culture brought with it the problem of how to handle many of our everyday problems directly ourselves. Our societies are structured around organizations whose role is to intervene between ordinary citizens and the spheres where the decisions that affect them are taken.

In politics, as well as other important areas like the provision of services or products or the creation of cultural content, there are organizations to whom we actively or passively delegate many of the decisions that are ours to take. For example, we leave it to the market and its organizations to decide on the best way to define the products and services we can access, and delegate the decisions that affect our lives as citizens to politicians. These intermediate organizations have different names, they are political parties, the media, multinational corporations, social organizations, etc.

The idea of disintermediation is attractive because it describes a scenario in which citizens, thanks to the lower cost of producing content generated by the Internet, are more and more (here is the observable regularity) equipped to compete with traditional mediating organizations. This thesis has been put forward and supported in one form or another by important researchers such as M. Castells (2009) or Y. Benkler (2006) and informs on key issues in the social sciences like social change and the digital society. In this book, despite the risk of losing our bearings like Fabrizo del Dongo, we will not question this assertion but try to review the effects that its advance has had on political communication and public opinion. Applying Thomas theorem, we assume that *if men define situations as real (disintermediation), they are real in their consequences.* In other words, if we as a society believe that Internet has arisen to change the rules of the game, we shall focus on what can happen as a consequence of this.

However, unlike what these authors suggest, we bet on a less radical version of disintermediation. That is, we call disintermediation a process of transformation of public communication in which traditional agents, now share space with new agents to which Web 2.0 has empowered. These new agents, in many cases citizens who are not in the public space in representation of structured organizations, disintermediate but also reintermediate. They disintermediate in so far as they reduce the weight of the traditional actors and re-intermediate as they themselves become central actors in the public space. However, unlike the industrial communication of which Benkler (2006) speaks, horizontal and Web 2.0-based communication allows, at least in power, any Internet user to become a mediator. This is the strength of Web 2.0 and, at the same time, the promise that legitimizes computer-mediated communication.

1.1 THE AIM OF THIS BOOK

As we said earlier, this book looks into the consequences of disintermediation. Specifically, the consequences of this process on public opinion. In other words, on the agents who act in the digital public sphere, mainly through digital social media. By choosing to study public opinion and not the digital public sphere itself, we will pay less attention to technological aspects (the structure of Internet and social media) and focus instead on the changes in the actions and behaviour of the agents.

In *The Structural Transformation of the Public Sphere*, Habermas (1962) establishes that the ideal of public opinion is that of an inclusive and autonomous system in which the agents are able to respond immediately and critically to any proposal. However, both the public sphere and public opinion itself[1] have experienced a range of different historical moments and, according to the German philosopher, they have never achieved this ideal standard that we have just outlined. Representative public opinion, in the early modern period, was a space in which the dominant political forces exercised their power of representation before a passive audience. Later, it was the bourgeois elite who added a critique of power and made public opinion rely on arguments and reasons. Even so, this public opinion, presented in the bourgeois public sphere (salons, bourgeois press, etc.) while proclaiming itself as the voice of the "public", meaning everyone, actually represented class interests. Those interests became legitimizing when bourgeois public opinion assumed the leading role in politics, economics and society.

---

[1] In opposition to the private sphere, the public sphere are those spaces of "social spontaneity free of state interference as well as the regulations of the market […]. These spaces for discussion and debate are public forums for reasoning, and it is from here that informal public opinion emerges, along with civic organizations and, in general, everything that questions, influences or offers critical assessments on politics" (Velasco, 2003, p. 170).

The standard form of public opinion sustained by free and equal citizens, based on rational debate, did not disappear during these periods. In fact, when the Internet began to expand in the main western countries, many academics and researchers started to discuss the effect this technology would have on public opinion.

Dahlgren (2005), for example, warned of the Internet's potential to upset the processes of political communications in representative democracies at the same time as he saw its potential to extend and make public opinion more diverse. Earlier, Papacharissi (2002) had sustained that Internet-based technologies would allow people in different parts of the world to talk to each other, but also that it would often fragment political discourse. The Internet and its associated technologies, according to Papacharissi, have created a new public space for politically oriented conversation, but whether this public space becomes an arena for exchanges between agents of public opinion or not does not depend on the technology itself.[2]

The emergence of Web 2.0 technology only served to stir up and enhance this debate. Major works by Benkler (2006) and Castells (2009) spoke of a radical process of transformation called disintermediation that affected, inter alia, political communication in the networked society. From this point of view, the main effect of this web technology was, as mentioned above, "disintermediation" or "horizontalization". In the realm of political communication, this implies that citizens will be able to, thanks to the lower cost of producing content for Internet, compete (although not replace) for public space with the organizations that have traditionally occupied it and mediated for the citizens and the general public (the media, political parties, unions, etc.).

In short, these authors and others showed us the way forward. The key idea here is that the Internet has created a space suitable for wider, more diverse public opinion (disintermediation). However, two questions arise from this. The first is whether disintermediation is really creating a new scenario; a scenario closer to the standard ideal described by Habermas (1962) which is inclusive and offers participants autonomy and the ability to respond critically.

---

[2]Our interpretation of Papacharissi's 'insinuation' is that whether Internet becomes a tool for innovating political communication does not depend on the technology but on the wider political and economic processes that can open or close opportunities for citizens to use this technology for their own purposes.

The second question is whether, as has occurred in the economic sphere, the system has generated ways of absorbing the disintermediation of digital public opinion or not, diminishing its innovative capacity. The disintermediation of public opinion was foreshadowed in the economic sphere. The Internet's well-known ability to reduce production costs meant that individual agents were able to engage in cooperation and collaboration to offer the market services that could rival the systems of industrial production. Well-known examples of this are: *Napster*, *Couchsurfing* and *Blablacar*. The economy, however, showed an impressive ability to adapt to the new scenario provided by Web 2.0 and came up with systems such as *Spotify* and *Airbnb* that, adopting the horizontal spirit of *Napster* or *Couchsurfing*, transformed them into traditional commercial tools.

Our paths diverge depending on the question we ask. We could be interested in how public opinion is structured in a digital context (question one) or in how the system transforms and changes it into a product that is market ready (question two). We have sketched some answers to the second question on markets that the reader can find in Robles and Córdoba-Hernández (2018) and Córdoba, Robles, Rodríguez, Cruz, and Semenzin (2017). In this book, however, we will examine the first of these points. We will study how the structure of public opinion in the context of digital political communication is marked by disintermediation.

To respond to this question, we shall start from the notion that the Internet in general, and Web 2.0 in particular, have lowered the cost of making content to the point where citizens can share the public sphere with the organizations that have traditionally dominated it. We are particularly interested in the effects of this process, which in general terms are threefold: First of all, disintermediation of the agents. This occurs when the leading political agents change their operations and/or their structure to adapt to the changing digital arena. They should also do this to actively include citizens. Some examples will illustrate this type of disintermediation. We have seen the emergence of a new model for political parties, the connected political party. This model is characterized by its use of electronic tools to outsource some of its internal processes and to put them in the hands of the citizens. This type of party, especially on the left, opens up processes such as the design of the electoral manifesto to citizen participation through open discussion.

In second place, we can identify a disintermediation process in messages, so that the traditionally powerful agents in the public sphere lose control of the production and publication of information. Thanks to the characteristics of the digital environment, the connection to platforms and technological tools, citizens become *prosumers* in the sense that not only do they consume messages, but also produce them on their personal devices. At the same time, by sharing text, images, videos, hashtags, posts, audios, etc. on social media, they also personalize the content by adapting it to their own circumstances, generating new messages and circulating them in different contexts.

Thirdly, as we shall see in the case of #BringBackOurGirls in Chapter 8, the disintermediation of public opinion can also happen in public spaces. Technological change has expanded the public arena. The visibility that certain messages, campaigns or causes can achieve through social media's the viral nature means they can go beyond their local context and reach a wider sector of public opinion. In other words, certain content, normally with great emotional power, is reproduced on a mass scale in different social and traditional media, jumping the barriers that existed previously.

In effect, the different social and political actors have changed the way they operate to adapt to the context of disintermediation. However, the question is whether disintermediated public opinion is critical, reactive and independent, in the sense of achieving the ideal raised by Habermas (1962). Does the context of disintermediation really create a scenario closer to the ideal for public opinion? The answer, as is always the case with big questions, is not straightforward. Various empirical studies into this line of research have returned polyhedric results.

This book will look into three cases, those of #UnidosPodemos, #Bring-BackOurGirls and #BlackLifesMatter, which represent the potential of the Internet to, respectively, make agents open up to the political culture of digital society (the culture of disintermediation), to generate and broadcast messages without the need for the traditional mediating agents and to generate global spaces in which citizens can harness support for their causes and thereby make them stronger. In short, these three case studies show the effect on individuals of the three basic consequences of the disintermediation of public opinion. That is, the disintermediation of the agents, the space and the messages.

However, does finding the patterns of disintermediation in individual's behaviour imply that we are necessarily dealing with critical, reasoned and independent public opinion? The famous engraving by Francisco de Goya says "the dreams of reason produces monsters", and the dreams of digital public opinion will do the same. Some of these monsters are related to the disintermediation of agents, spaces and messages and, others, such

as polarization or incivility, are part of the same structure of digital com-
munication. We can see them clearly in the case studies that form part of
Chapter 9 of this book, in the 2016 presidential campaigns in both the
United States and Spain.

Polarization is, to a certain extent, a fairly predictable phenomenon in
two-party contexts where public opinions, disintermediated or not, must
take the side of one or the other of the political parties' candidates. The
problem of polarization runs parallel with other political processes that,
like populism and *negative partisanship*, tend towards extremes and the
distancing of parties instead of convergence at the political centre. It is in
these cases that polarization becomes a problem for democracy and a public
opinion based on reasoning and debate.

Incivility, or the use of rude expressions with the aim of insulting or
upsetting the other person is another of the greatest risks to the proper
conduct of the digital public opinion. As we saw in the case of the 2016 elections in the United States, hatred of the other candidate and the expression
of this stance are obstacles to reaching agreements or a certain amount of
common ground, yet it is constantly increasing in debates on social media.
The same thing happens when the main objective is to inflame the conversation. This is not just the work of human agents, but the contribution of
machines designed to generate noise in public debate and provoke angry,
outraged responses which of course distance the debate even further from
the ideal for communication.

Even do, our studies show that this type of behaviour is more frequent
in accounts on very active social networks. In other words, accounts that
potentially belong to large organizations with a high capacity to put a
huge number of messages in circulation. The people responsible for less
active accounts that share only a few messages, however, show less polarized
behaviour and are less inclined to reject others. This means that if we were
in a fully disintermediated scenario, in which individual agents had the
same weight as the traditional agents of public opinion, our results indicate
that the situation would be less extreme and adversarial. Reality is, however,
rather different. Negative partisanship, incivility and flaming are ubiquitous
in public debate and distance it from the ideal that we all aspire to, essentially
due to the importance retained by traditional political actors.

We consider that, although the effects of disintermediation are enormous for political communication, the monsters of communicative reason
obscure, if not dilute, the possibility of a critical, argumentative and independent public opinion. That is, an ideal public opinion as it is described

in the work of central figures such as H. Habermas (1962). It is difficult to live with the spirit of understanding, openness and transparency when the debate is filled with people who only seek to blow up the patience of the opposite with false or provocative news. It is impossible to deepen consensus from the reinforcement of the starting positions generated by polarization. It is more than risky to suppose that the inclusion of citizenship and horizontality can persist in a space in which insult and stereotype are the usual way of expelling the others from the debate.

The question is, while the disintermediation of agents, messages and spaces open spaces to dream of a more inclusive and rational public space, the phenomena arising from the first steps of this dream, make us think of nightmares. Thus, in the face of the question, really, does the context of disintermediation generate a scenario closer to the ideal of public opinion? The answer is clear; we are potentially in a position to achieve it while, in practice, we do nothing but get away.

In the empirical sections of this book we have taken data from digital social networks, mainly Twitter, on the assumption that it is one of the most representative spaces in the digital public arena. Case studies from Spain, the United States and Nigeria will be examined, but all of them have become global via the Internet. In the analysis we have used big data techniques such as *sentiment analysis*, *social network analysis*, *data mining* and *Computer-assisted learning*. We also used other commercial APIs to discover the behaviour of hashtags and influencers, such as: *Topsy*, *TalkWalker*, *Twitonomy*, *TweetArchivist*, *CartoDB* and *Hastagify.me*. This work, given its nature and audience, will place special emphasis on the results and less on the techniques employed.

The book is based on prior studies made by authors with their corresponding research teams, so interested readers can find all the technical explanations about the methods used in those publications, which will be conveniently indicated in each case.

References

Benkler, Y. (2006). *The wealth of networks: How social production transforms markets and freedom*. New Haven, CT: Yale University Press.

Castells, M. (2009). *Comunicación y Poder*. Madrid, Spain: Alianza Editorial.

Córdoba, A. M., Robles, J. M., Rodríguez, A., Cruz, M., & Semenzin, S. (2017). Movimientos en (des-) acuerdo con la Red ¿Mercantilizando o haciendo común Internet? en R. Cotarelo & J. Gil (Comps.) *CIBERPOLÍTICA: Gobierno abierto, redes, deliberación, democracia* (pp. 219–237). Madrid, Spain: Instituto Nacional de Administración Pública.

Dalhgren, P. (2005). The Internet, public spheres, and political communication: Dispersion and deliberation. *Political Communication, 22*(2), 147–162.

Habermas, J. (1962). *Strukturwandel der Öffentlichkeit. Untersuchungen zu einer Kategorie der bürgerlichen Gesellschaft.* Frankfurt, Germany: Suhrkamp.

Papacharissi, Z. (2002). The virtual sphere: The Internet as a public sphere. *New Media & Society, 4*(1), 9–27.

Robles, J. M., & Córdoba-Hernández, A. M. (2018). Commodification and digital political participation: The "15-M Movement" and the collectivization of the Internet. *Palabra Clave, 21*(4), 992–1022.

Velasco, J. C. (2003). *Para leer a Habermas.* Madrid, Spain: Alianza Editorial.

PART I

The Framework: Towards a Disintermediated Politics?

In 1999, Shawn Fanning and Sean Parker, who were barely 19 and 20 years of age, created *Napster*, the first peer-to-peer (P2P) file sharing programme, in order to share music on the Internet, so that people all over the world could access their favourite group's catalogue without having to pay for it. For most of us *digital immigrants*, the appearance of *Napster* meant the end of an activity that verged on a sacred ritual; endlessly browsing record shops in search of discs that were high on our most wanted lists. We were, however, willing to forego this pleasure in exchange for the ability to hear everything we wanted without paying for it.

The rise of Napster quickly provoked the anger of musicians and copyright associations such as the *Recording Industry Association of America*, who decided to take legal action against the company. The pressure and financial penalties imposed by the courts of North America forced Napster to file for bankruptcy in 2002, after which it was bought and sold by the music industry until its business model changed to what it is today: one of the largest stores of legal music on the Internet, in competition with Deezer and Spotify.

Despite this transformation, we can say that the platform and those that came afterwards changed many of the paradigms of the music business, forcing artists to diversify their revenue streams to maintain their standard of living. The album sales in record stores were no longer enough, and artists had to get out and charm their fans on stages around the world.

Five years later, the revolution would come to another sector. In 2004, a non-profit platform started up in San Francisco called *Couchsurfing International Inc.*, and it offered its users the chance to exchange accommodation and services on the social networks, an idea that set the entire traditional hotel system trembling. On this website, travellers contacted others, who they did not necessarily know, in the city that they wanted to visit. These people made their sofa available for the night in exchange for similar services in their home town. This made travel a very different proposition, much more amateur and flexible, and together with the growth of low-cost airlines, radically cheaper, so that many people with limited means could now get to visit their favourite cities.

From this point on, young and not-so-young people began to coordinate with each other on the Couchsurfing platform that, in 2011, already had 3 million members. However, like Napster, after dodging the pressure and reaction from the hotel sector, it entered a process of transformation from which it emerged in a standard company format.

Peer-to-peer phenomena, such as the exchange of music files or accommodation between Internet users, angered the mainstays of the traditional economy, who responded with three different strategies. First of all, many of the activities that were beginning to be accepted as normal between users, such as P2P file exchange or low-cost travel between cities, were suddenly declared illegal or treated as acts of piracy. Secondly, governments around the world began to change their laws to provide protection for digital content from any non-commercial use. Finally, the companies in the sector affected by these changes began to offer services that, while still costing money, gave the appearance of accepting the banned services' spirit of "collaboration" and "coordination".

Leaving aside the regulatory context and their legal and fiscal aspects, cases such as Napster and Couchsurfing are just examples of initiatives that rely on networks of collaboration. It is a very long list and includes famous cases like AirBnB and Uber which, like other startups, have expanded and consolidated their position as disruptive business models in different productive sectors. They are all examples of how web 2.0 has made it possible for citizens to start creating an alternative digital economy in which mediators are noticeably absent.

Who are these mediators? Naturally, they are organizations, in this case companies, who offered the service that all citizens used before these collaborative tools appeared on the Internet. Prior to Napster, the record industry provided us with services, in the form of vinyl LPs and later CDs that we purchased. Before Couchsurfing, the hotel trade set the prices and options for accommodation that we could choose from, and there were telephone operators who took our calls and instructed a taxi driver by radio to come and pick us up.

What is the phenomenon that lies behind all this? As Keen (2015) said, we are living in a major economic, cultural and intellectual transformation in which we are all connected. As McLuhan accurately predicted, digital communication media would spark a revolution as radical as the Gutenberg printing press did in its day. This is because it has replaced the linear, vertical technology of industrial society with an electronic network shaped by circuits of continuous information.

In short, the revolution of the twenty-first century is a technological one with potentially quantifiable results in which every aspect of today's world (education, politics, the economy, transport, health and the financial sector, etc.) is subject to processes of radical disruption and is forced to rethink the way it operates. What drives this scenario is citizens' ability to engage in exchange and collaboration, the central strategies of a new type of interaction that are marked by lower costs and the handling of web 2.0 tools.

As in the economy in general, P2P business has transformed the nature of the market and its systems, so the question addressed in this book is how this transformation and the boom in digital connectivity tools has affected the processes of political communication between citizens. To what extent can we talk of a real change in the way that citizens debate political issues? To what extent are new strategies emerging to influence those in power in relation to the issues that affect us? As in the economy, is there a process which reduces the importance of mediators? How have the main political agents responded to this situation? What measures has the media taken to retain its function of mediation?

The arguments presented by experts like Castells (2009) say that we are facing a radical transformation in the processes that produce political content. The digital revolution has led to a remarkable repositioning of citizens, who the Internet has allowed to take a leading role at the expense of the actors who have traditionally been at the centre: political parties, mass media, structured social movements, etc.

Many authors claim that disintermediation is also extending into the area of public opinion. Castells (2009), for example, talks of the horizontalization of political communication that enables citizens to occupy a prominent place in contemporary political processes. Benkler (2006) offers a similar interpretation when discussing the freedom that social media offers citizens interested in politics, which is comparable to what has happened in the economy.

As we will try to show in this book, disintermediation, neither economic nor communicative, is a process that can be measured in black or white terms. The technologies do not arrive from one day to the next and transform the mediation relations that have characterized the social, political and communicative structure since the industrial revolution. However, we do observe that there is a process of disintermediation in the style defined by Castells or Benkler that is characterized by the emergence of new actors (mainly citizens) who share the functions of social mediation with organizations that once monopolized that function. Thus, in the economic and services field, along with regulated musical education (academies, conservatories, tutors, etc.) we find amateur citizens who offer free music classes through YouTube. The traditional mediator between the musical knowledge and the student persists, but must coexist with a new agent, the amateur teacher. This, thanks to having a technology that allows you to record quality videos with little cost and hang them on a global and free distribution platform, is part of the new reality of music services.

This structure can be transferred to the political communication in the digital public space. In the same way that service companies, political parties or the media are seeing how, with a mobile phone in hand, a non-professional citizen, can produce content that can be transmitted virally to a potentially huge number of people. That and many other citizens are disintermediating as they struggle with traditional mediators for the control and definition of the public agenda. In doing so, they become new mediators or "re-mediators" whose main characteristic is that they are not professionals of political communication and public opinion.

So, economic disintermediation had a measurable effect on certain patterns of economic behaviour and consumption. These changes led to a scenario in which coordination and collaboration emerged as central values. Even so, this transformation was tempered as a result of the economy's ability to adapt to these practices. The issue this book deals with has parallels with this as we ask: What changes has the disintermediation of public opinion brought about? Which political actors are affected? What reactions, if any, have been generated in the political system?

To answer these questions, we have to begin, as we said early, with the concepts of public opinion and political communication. Social sciences have shown that these concepts are far from clear, and any definition will generate intense debate about their condition. We shall, however, accept this ambiguity and try to use these concepts effectively to lay the groundwork that will enable us to provide answers to our research goals.

REFERENCES

Benkler, Y. (2006). *The wealth of networks: How social production transforms markets and freedom.* New Haven, CT: Yale University Press.

Castells, M. (2009). *Comunicación y Poder.* Madrid, Spain: Alianza Editorial.

Keen, A. (2015). *The internet is not the answer.* New York, NY: Atlantic Monthly Press.

CHAPTER 2

The Mediated Public Opinion: When Everything Happens Through Others

Political communication has always existed. Ever since an idea of public opinion has been around, that is. In other words, since a person or group of people or a whole society has shared their opinions on a subject of common interest. Defining what public opinion is, however, is no simple task. Following Kant, many of the most outstanding social scientists have dedicated time and energy to resolving this issue. In general terms, however, public opinion is understood as the tendency, or preference, whether genuine or stimulated, of a society or an individual on social issues that are of interest.

Our goal in this first chapter is to provide a context for our study of political communication and public opinion in the Network Society. We are particularly interested in a historical perspective that not only specifies the forms that public opinion and political communication have adopted in different periods, but that also identifies the actors and their motives for leading public opinion in each of these periods.

We want to start by taking the work of Habermas, *The Structural Transformation of the Public Sphere* (1962): *An Enquiry into a Category of Bourgeois Society* as a reference, but instead of debating this proposition, we want to identify the historical milestones he mentions that we will then use in the rest of the book. Therefore, to a large extent, this section will be a partial and subjective review of this work and will adopt its main ideas as inspiration.

© The Author(s) 2019   17
J. M. Robles-Morales and A. M. Córdoba-Hernández,
Digital Political Participation, Social Networks and Big Data,
https://doi.org/10.1007/978-3-030-27757-4_2

2.1   The Idea of the Public Sphere: A Rereading of Habermas

Public opinion is popularly understood to mean "expression of opinions". Public opinion surveys therefore focus essentially on the measurement of the stance that citizens adopt in relation to a long list of topics of greater or lesser significance. This aspect, however, despite the widespread acceptance of this meaning and the solid results it offers for understanding social behaviour, does not match the concept of the public sphere that we will be working with here. The expression of opinions, that is a feature of our society, is one more outcome arising from the concept of public opinion's loss of meaning that affects the critical and controlling function that it ought to have (Habermas, 1962).

In order to understand our interpretation of the concept of public opinion, we must first underline its contingent nature. In other words, it is defined by the forms adopted by the social structure. It is useful, therefore, to "label" public opinion in order to specify which context of public opinion we are referring to at any time. In this sense, Habermas (1962) uses the expression *representative publicity*, the first type of public opinion that we will discuss, to refer to forms of public communication that were used prior to the modern period.

From the fifteenth century onwards, Europe embarked on the development and consolidation of a new form of government that was characterized by cultural and territorial unification under a strong monarchy. With a few notable exceptions, such as the case of the republics in the north of Italy, monarchs such as Isabel I of Castile and Fernando II of Aragon, or Louis XII of France represent this idea of restructuring the state and the social organization of their kingdoms.

In December 1492, the same year in which the inhabitants on both sides of the Atlantic came face to face for the first time, the Catholic Monarchs, Isabella and Ferdinand captured the city of Granada from King Boabdil. Although the Muslim presence in Spain did not end with the victory over the troops of Boabdil (the expulsion of the *Moriscos*, the inhabitants of the Iberian Peninsula of Muslim origin who had been baptized and converted to Catholicism, did not end until 1609), the defeat meant the final loss of political power that this culture's representatives had enjoyed in Spain since the year 711.

As a result, and in recognition of their struggle against the Muslim enemy, Pope Innocence VIII bestowed the title of Catholic Monarchs on

Isabella I and Ferdinand II. From this moment on, the figures of Isabella and Ferdinand not only *publicly represented* their kingdoms of Castile and Aragon, they also acquired the status of *representatives* of the Catholic faith in the different kingdoms that became modern Spain, Portugal and their American colonies.

Isabella I and Ferdinand II are, from our point of view, examples of this first type of publicity. Representative publicity does not imply just any way of making ideas or motives public, but rather a certain kind of status that is conferred, in the case of the Catholic Monarchs, by their condition of being monarchs (that is, by birth) and by being granted the title of "catholic" by Innocence VIII in recognition of their merits. When they appeared before their subjects, those subjects would adopt the role of a passive *audience* for the monarchs' public aspect. This means that Isabel I and Fernando II represented all their values and authority, every time they appeared in public. In this sense, the symbols, emblems and regalia are all tools that contribute to the publicity to reinforce the image of this type of representation.

Louis XII could be taken as another key example. To strengthen his position as King of France and as part of his reorganization of the structure of the state, the king embarked on a public representation strategy that aimed to show his subjects two sides of this nature: one as a Christian and the other profane. As regards the first of these, Louis XII portrayed himself as a good shepherd who would give his life for his flock (in a clear reference to the Shepherd that the Catholic church associated with God, and to his people as a flock of sheep, an accepted example of passivity in need of protection). On the other hand, to maintain his image as a statesman and political leader, the French king revived the iconography of imperial Rome and presented himself in public as the heir to Charlemagne.

The public assertion of these roles is one of the pioneering forms of publicity. *Representative publicity* is a public expression of dominion, of a state of affairs that was ruled over and exercised by feudal lords and the kings who were behind this new model of western monarchies. In this sense, the sovereign and the institutions embody the country (he is the law) and exercise their dominion "for" the people and "before" the people. It is a sign of social status, that is placed before a passive audience for the aura or "grace" of nobility.

This form of publicity is particularly important because, according to Habermas, it was inherited by western societies in the earliest stages of capitalism. Whereas the forms of courtly life used to be displayed for the people, gathered as a contemplative, acquiescent audience for the majesty of the prince in palace balls, in the early stages of capitalism this audience was fed stories by the periodical newspapers that were subject to State control and censorship. In both cases, active participation was not possible for the majority of the population.

This was the context in which the second concept in the historical development of the public sphere emerged, that of the *bourgeois public sphere*. The growth of capitalism and the economic repositioning of the bourgeoisie saw them reappraise their condition as an audience and with it the need to develop a greater capacity to respond to and oppose the State. The *bourgeois public sphere* arose from this situation as a new public space. Habermas saw the proliferation of bourgeois salons, with their hosting of critical debate, as the expression of the emergence of this new public space in which the bourgeoisie began to engage in an inclusive critical discussion of the state. This was the start of a process of "self-understanding" on the part of a "reasoning audience" which has to dialectically resolve the antithesis created by budding capitalism between the private sphere, where most of the population found itself, and the public sphere, which had for the most part been seized by the State.

One important characteristic of this new scenario for public debate was that citizens who took part in the bourgeois public sphere were informed and educated about the issues of interest through the debates in the new bourgeois newspapers. These debates also came to acquire the form of rationally convincing arguments. This was a fundamental difference compared to the representative publicity in which the state controlled everything that became public, as well as who that audience were going to be. In short, the bourgeois public sphere is marked by the presence of educated actors who were able to use rational arguments to back up their opinions. One of the key arenas for this meeting of minds was the bourgeois newspaper.

In the case of Spain, which we used as an example of the previous form of publicity, this process arrived late, but found clear expression. The widespread appearance of the press in this country can be dated to 1734, although there were a few precedents. This same press, though, was always controlled and censored by *El Consejo de Castilla*. (The Spanish monarchy's main body of power from the sixteenth to the nineteenth century.) There was a certain degree of independence in cultural and financial issues (both traditionally reserved to the private sphere) but political matters were controlled by the monarchy to the extent that we cannot truly speak of a bourgeois public sphere in Spain during the eighteenth century.

The control of the press became even more rigorous in the wake of the execution of the French royal family in 1793. This historical event of enormous significance sowed uncertainty and fear in the Spanish crown and stifled the emergence of a bourgeois public opinion until the royal family left the country as a result of the Napoleonic invasion and, in an atmosphere dominated by bourgeois salons and round tables, the middle classes began to adopt political positions and to see the press as an arena free from interference. This is why the decree for a free press was not signed in Spain until 26 October 1811.

The expansion of the *bourgeois public sphere* brought with it two processes that are features of modernity. The first of these was political activity and debate, which was increasingly subject to rational rules or regulations that originated from a public opinion that took up a position in opposition to power. Here we can see another difference with respect to *representative publicity*. While the latter is guided by the acts that obey the will of the sovereign, the bourgeoisie relied on reasoning and arguments. This is a transformation that, although stereotyped and devoid of meaning, remains a central part of the current concept of public opinion.

The second process is characterized by the capacity of this type of public opinion, the *bourgeois public sphere*, to be identified with popular public opinion. In this way, an opinion that represents a particular segment of the population, the enlightened bourgeoisie, acquired hegemonic status. Therefore, in this new context at the close of the eighteenth century and start of the nineteenth, public opinion was liberated and sought to distance itself critically from the state. However, at the same time, this independence came about at the expense of the privatization of the public sphere. In other words, what was considered "public" in the notion of "public opinion" was really the private interests of the new bourgeois class.

This was the first of the fundamental contradictions that Habermas presented, according to our interpretation. When public opinion ceases to be heteronomous, it begins to be privatized. This new disposition of things, however, leads to a second contradiction. Although it springs from private interests, the *bourgeois public sphere* is a critical counterweight to the power of the state.

This double paradox reached even greater heights in the twentieth century, when Habermas was writing his work. As the author noted, the accumulation of capital and legal power meant that this *bourgeois public sphere* began to veer away from its critical posture and towards one where it began to justify the existence of the "status quo". In other words, the *bourgeois public sphere* continued to control this public space, now through the mass media, but abandoned its critical function to become a tool of legitimation.

In these circumstances, the *bourgeois public sphere* was transformed into an uncritical and legitimizing sphere. That is, a public sphere that until recently denied access to large sections of the population and fought, at the same time, to maintain its position. Although at certain points of history it displayed positive traits, especially in its critical dimension with *representative publicity*, and negative traits, this desire to pass off its class interests as the public interest throughout the twentieth century outweighs the former and preserves only the latter.

2.2   THE NORMATIVE CONCEPTION OF PUBLIC OPINION

Not only does Habermas explain the nature of this process that we have interpreted here, but he offers a synthesis of the aspects that, in his view, should define public opinion as accepted norms. Following C. W. Mills (1956), he claims that public opinion should have the following characteristics. First of all, it should work so that there are as many people expressing opinions as there are people to receive the opinions of others. This means that political communication is not dictated by one (the monarch under *representative publicity*) or a few (under the *bourgeois public sphere*) but is open and therefore truly public.

Secondly, political communication should be sufficiently ordered so as to respond effectively and immediately to any opinion publicly expressed. In other words, it should have a structure that allows public opinion to be created efficiently and, at the same time, should consist of a system of debate and the rational contrasting of ideas.

Third, political communication should offer an easy and effective response in opposition, where necessary, to the system of authority. According to our interpretation, this means that it must be critical whenever the circumstances of reasoned debate call for it. Finally, authoritarian institutions should not infringe on the public space so that it can be more or less autonomous in its activities.

However, we are apparently seeing the opposite as public space seems to be organized so that only a specific group of people are in a position to express their opinions in relation to the information they receive (the public, in general, is a group of abstract beings who obtain information from the media). The prevailing system of communications is so impermeable in its organization that it is impossible to offer an effective and immediate response. The channels that enable opinion to be transformed into action are organized and controlled by social, political and financial elites. Citizens do not have autonomy with regard to institutions. According to Habermas, in the middle of the twentieth century, we would be seeing a return to *representative publicity*.

When Habermas wrote *The Structural Transformation of the Public Sphere*, it was difficult to see real public opinion, or at least a version that was similar to the one presented in the latter part of his work. Whether due to the absence of public opinion (*representative publicity*), or its partial nature (*bourgeois public sphere*) or the loss of the critical dimension of the bourgeois public sphere (representative publicity in the context of the mass media), our society has always suffered from a deficit in the context of its public sphere. However, as we shall now see, the rise of Internet and Social Media has driven many important authors to reassess this question. We would therefore like to present the key elements of public opinion in the digital society.

References

Habermas, J. (1962). *Strukturwandel der Öffentlichkeit. Untersuchungen zu einer Kategorie der bürgerlichen Gesellschaft*. Frankfurt, Germany: Suhrkamp.

Mills, C. W. (1956). *The power elite*. New York, NY: Oxford University Press.

CHAPTER 3

The Culture of Politics on the Network

Mediation in the sphere of public opinion was dominated, prior to the arrival of Internet, by the mass media and political parties. The former has operated, as shown in the abundant literature on the subject, as the controllers of the agenda for the core topics of public opinion, and as the agents who define the reality that we can debate. In this sense, political communication takes the form of a one-directional channel in which the public is an agent that consumes what others publish.

Political agents and parties adopted a more or less closed organizational structure in which the messages and content expressed in public were defined internally beforehand. The mechanisms of representation allowed citizens to indicate their positions to the parties without having any direct influence on the decisions taken or the design of the road map that the party would follow.

It was these structures for producing content (media and party structures) that have been modified as a result of disintermediation. In this section, and before we examine the effects of disintermediation, we will offer a brief definition of the starting conditions (for the parties and the media). In point 2.2, however, we shall try to define the concept of disintermediation in order to contrast the political culture of the social networks with the political culture of mass public opinion.

© The Author(s) 2019
J. M. Robles-Morales and A. M. Córdoba-Hernández,
Digital Political Participation, Social Networks and Big Data,
https://doi.org/10.1007/978-3-030-27757-4_3

3.1   A CHANGE IN PERSPECTIVE

Our understanding of democracy rests on three fundamental supports: sovereignty, the separation of powers and political representation. This last concept is subject to a far-reaching process of transformation as a result of the technological revolution. This transformation affects, as we shall try to explain here, the forming of public opinion.

When we talk about representation, we are referring to the function that enables decisions that affect all citizens to be taken without their direct involvement. We could say that the standard view of representation is the transfer of power to decide which political actions should be carried out. In a democracy, this transfer is embodied in the vote that citizens use to select one candidate over another and entails a certain degree of reciprocity, rendering of accounts and the possibility of dismissal should candidates fail to honour their obligations (Sartori, 2005). This reciprocal understanding is based on awareness of the preferences and opinions of citizens which are expressed in public (public opinion).

This concept, however, is difficult to define. In his classic work, Pitkin (1967) showed how political representation can be understood in five different forms: (a) as *authorization*, when the representative is someone qualified to act on behalf of the represented; (b) as *rendering of accounts*, to the extent that the representatives have to account for their actions to the citizens; (c) as *descriptive* representation when there is a correlation of shared qualities or characteristics between representatives and represented, so that they are the "reflection", "image" or "mirror" of those who selected them; (d) *symbolic* representation when the representatives have an emotional identification with those they represent and (e) *substantive* representation when the representatives are concerned mainly with the benefits and interests of the group of electors.

Later works by Mansbridge in 2003 and 2011 seek to delve deeper into the meaning of "representation", a concept that is central to our political co-existence. For this author, the *quid* lies in the fulfilment or non-fulfilment of the promises made by the representatives, so that voters can choose to punish them or support them in the next elections or replace them before their mandate has expired. From this point of view, Mansbridge introduces the idea of Surrogate representation, in which representatives assume the power to embody the will of those who did not directly select them.

Representation, as we have shown, is a complex issue, with different dimensions that co-exist and overlap in practice (Rehfeld, 2011). What the above authors do agree on, however, is that representation includes some form of reciprocity or affinity between the representative and those represented, either through promises, anticipated preferences or affective identification, so that the voters can exercise some measure of control. Citizens therefore delegate the power to decide which political actions should be taken to their representatives. Representation can be seen, whether as reciprocity or affinity, as a form of mediation between the publicly expressed interests and preferences of the citizens and the implementation of measures that express these preferences. This is the point at which public opinion becomes a key element for representative democracy.

Especially since the decade of the sixties of the last century, other political actors such as social movements or trade unions have become part of this system of political communication. These actors complement or function as a critical counterweight to the opinions and points of view offered by political parties and institutional actors. In this way, and although political parties continue to have a hegemonic position in the public political debate, collectives such as social movements or other civil organizations have become central mediators on increasingly diverse topics.

In most cases, however, the link between the political decisions made and the will of the citizens as expressed in public opinion is not only mediated through the action of their representatives, but also through the media. To a large extent, citizens also delegate the responsibility of acting as a window on the world around them to the media. It is no accident that classic studies of political communication acknowledge the transcendental role of the media by calling it the "fourth power". Like their representatives, the media are subjects that have been granted legitimacy by the populace to speak in the public sphere.

In the political structure of power sharing, prior to the appearance of Web 2.0, the "media logic" referred to by Altheide and Snow (1979) was considered to be a form of communication that sustained a particular interpretation of the world that was subject to organizational and institutional factors. In other words, these authors said that understanding the media logic was essential in order to grasp how these agents affected public opinion in its inclusive aspect.

Since its emergence, mass media has been characterized by its capacity to produce expensive, highly professional content. This mass media generated and selected the content that would then be broadcast unilaterally to the general public. This is why, as Klinger and Svensson (2015) said, mass communication has been vital to the creation of a public sphere, founded on the experience of participating in collective communication.

Although the *hypodermic needle* theory of Harold Lasswell (1927) may seem outdated to us now and has lost much of its original meaning, it is true that until recently, classic mass media was entrusted with delivering messages. To some extent, they acted as filters and decided what content could be published and broadcast. This idea originally came from the theory of *Agenda-Setting* in reference to the media's role in deciding which issues are more or less important for the audience (and by extension, in public opinion), thanks to the power they possess to focus citizen's attention on specific political topics and problems (McCombs & Shaw, 1972).

The media do not just select the topics, however, but also offer a certain viewpoint (Entman, 1993; López-Escobar, Llamas, McCombs, & Lennon, 1998) within a context that is clearly intertwined with the political system (Hallin & Mancini, 2004). The media "mediates" between events and their readership that is not actually present at these events. They open "windows on the world" through which citizens can see themselves and others and the political situation in which they live (Tuchman, 1978). The media is not always neutral in this role of mediating or "mediatization" of reality (Strömbäck, 2008), because they choose topics for a reason, delegating the autonomous construction of their political-electoral preferences to their audience.

As a complement to this view of the mass media as intermediaries, we can look at the approach taken by Hjarvard, Mortensen, and Eskjær (2015), based on the metaphors suggested by Meyrowitz (1993), on the media as channels, as language and as environments.

In the first of these, the media as channels, their main role is to transport messages from the issuer to the receiver. Our attention focuses on the technical ability of the media to spread the content in time and space. Seen this way, we can say that the media is essentially "an amplifier" that can raise the volume of an event.

When we consider the media as a language, however, what takes precedence is its capacity to shape messages and to display events to the public. This metaphor sends us back to the theory of *Framing*, (Entman, 1993), in which journalists seek to "shape" the events in a certain way so they can be inserted in a specific frame of reference and thereby give a story a certain "angle".

The third metaphor looks at the media as an environment, in the sense that it places them in perspective as part of a media system that is integrated with society and culture so that they are a feature of the environment in which people live and act. This is an interesting view because it defends the idea that the media, as part of an environment or system, contributes to the structuring of the ways in which people interact and relate to each other. For example, the social elite usually has better access to journalists, and the media also help to broadcast the discourse of these sources by granting them credibility and offering them a channel through which they can express themselves and be heard.

In this double context (representative and mediated), public opinion is a patient and heteronomous object on two levels. First of all, because it is seen by the representative political system as expressing public views. In other words, it is subject to scrutiny by those who wish to see (measure) citizens' views on a range of topics. However, it is a public opinion which is not subject to reasoned, critical debate, but rather "recreated" from questions that are asked individually to citizens through instruments such as opinion polls. Secondly, and as a complement to the first, this public opinion is mediatized by the structure of media production. Not only because this structure defines the topics of public debate and is thereby guided by the interests of a select group, but because it defines problems and reality itself in a way that projects a certain inclination about politics and the world in general.

In the first decades of the twenty-first century, however, the potential of interactive tools has opened up the debate about the relations between political parties, the media and citizens. We are living in a period that some authors have called Post-mediatic in which, referencing Marshall McLuhan's iconic phrase "the media is the message", the "who" is no longer so important as "what" is said (Jiménez, 2014).

Political parties, civil organizations, social movements and the media are positioned, thus, as the main agents of the public space of mass society. On the one hand, political parties, their leaders and political representatives publicly express their positions on issues of political relevance, so that, at least potentially, they are publicly debated or become a source of information so that citizens can accept or reject the government action of the different parties. For its part, the media contextualize, frame and define what should be a topic of debate in the public space offering a particular and sometimes biased interpretation of this debate.

It is precisely this state of affairs that will be altered by the popularization of digital technologies and the development of Web 2.0 tools. The thesis of disintermediation, which we will see in more depth below, proposes a transformation that not only affects who the mediators (media and political parties) are in the public space that debates political issues, but also the type of messages that are issued, their context of emission and, even, the way in which the issuers are organized. It is, if we follow this thesis, a profound transformation of communication.

3.2   The Richness of the Networks

Throughout the last decade, experts studying the Internet have noted the emergence of a new digital phenomenon, although they may have given it different names: *digital production* (Schradie, 2011), *peer production* (Benkler, 2006), *digital participation* (Lutz, Hoffman, & Meckel, 2014), etc., that exploit the possibilities that digital technologies offer to produce content for sharing on a global scale. This type of content may deal with finances, tourism, culture, politics, etc. However, one of the most outstanding common elements in this use of the Internet, according to the literature, is that it contests, or at least challenges the vertical production structures of industrial societies that preceded the information society (Benkler, 2006) and empowers the citizens who make it (Lutz et al., 2014).

The idea of the *information society* arose from the social and economic processes disrupted by communication technologies at the end of the twentieth century, defined by Castells (1996) as *informationalism*. In this context, the creation, processing and transmission of information are sources of production and power in society.

This concept came to complement that of the *knowledge society* (Bell, 1973, 1976; Knorr Cetina, 1998; Stehr, 1994), which emphasized the role of technology in obtaining knowledge and the relation between this process and the rising levels of learning and education among citizens.

In Castell's (1977) original view, and that of the theorists of the knowledge society, organizations, whether political, social or financial, continue to play a central role in the structure of society. For Castells, companies, social movements, political parties or the states themselves become central agents of society insofar as technology allows them to administer information.

However, in recent years some theorists have moved away from this idea of the central role of organizations and adopted another view in which citizens become protagonists in the management and processing of information and knowledge. Benkler (2006) holds that technological tools, collectively referred to as Web 2.0, especially social networks, enable the "citizen-amateur" to carry out professional activities that were previously reserved for organizations, from the production of goods and services, as we saw in the introduction in the first part of the book, to the creation of political and cultural content. In this sense, we could talk of a public opinion with the capacity and resources to generate its own dynamics for discussions and responses.

The transition from organizations to individuals is linked to technological factors such as: the democratization of access to personal computers, the extension of the Internet to different social classes and the appearance of tools that disrupt the traditional one-way traffic from producer to consumer, such as social media, and other social dynamics encouraged by the Internet (Benkler, 2006). Whatever the means, this process of individualization has been the central axis of various studies that, since the start of sociology as a discipline, have sought to describe the cultural changes that are affecting western societies (Zabludovsky, 2013).

The process we are describing originates in the area of economics. In *The Wealth of Networks*, Benkler warned of an unknown phenomenon that showed that Internet was growing by increasing the freedom and autonomy of citizens while shifting the balance of power between the market and the State towards civil society. He claimed that the key to this transformation is that the economy that would result from these changes would be based on collaboration and non-ownership, unlike traditional models. In his view, the reduction of production and organizational costs enabled by the digital tools of Web 2.0 would allow citizens to produce goods without expecting material returns or copyright. The ease with which these digital services can be coordinated and used are powering a collaborative system on a global scale.

This economic transformation would become the base of a process of social change that affects different areas such as: the culture and value system (Jenkins, 2006), information (Sampedro, 2014) or politics, and this is precisely where our research is focused.

Out of all these areas being transformed, as we have pointed out, one of the most noticeable has been the production of content for communication, and therefore for public opinion. As Castells (2012) explained, digital tools are at the heart of *mass self-communication*, with which any individual can broadcast a personal message on the Internet and achieve local, national or global resonance. What we can see in global protests, such as those of Tunisia, Iceland, Egypt, Spain, the United States, Turkey or Brazil in recent years, is how quickly political messages are being changed by the digital technologies that make it possible to create a basic mechanism for power in the networked society: *the power of interconnection* (Castells, 2012).

In this sense, Bennett and Segerberg (2013) indicate the effect of interconnected protest actions and social criticism. For these authors, one of the consequences of Web 2.0 is the transformation of the logic of *collective action* (Olson, 1978) into a logic of *connective action* in which citizens use social networks to organize and coordinate. It is a phenomenon of the personalization of politics through digital media in which people look for more flexibility to associate themselves with causes, ideas and organizations through individual action. *Connective action* does not require many organizational resources nor strong collective identity (Bennett & Segerberg, 2013). Here is where the Internet would be diminishing the importance that organization and identity has traditionally had as part of the explanation for citizen political communication.

The other key element in the explanation offered by these authors is the emergence of a new logic of *connective action* that is based on sharing personal content on social media (Bennett & Segerberg, 2013). As a result, the Internet is becoming the sphere of political socialization in which citizens produce and/or distribute content to express their support for certain causes or to question others. They point out the emergence of public opinion organized according to the logic of the network, being horizontal, de-structured and dominated by immediacy.

This interpretation of the effects of the Web 2.0 on politics is not only limited to the sphere of public opinion but also extends to conventional political activity. Gibson and Ward (2009) apply the term *disintermediation* to the process by which Internet makes it possible to counter the weight of the political organizations that have traditionally wielded greater influence and rallying power through the emergence of groups of citizens who have organized through the network (Wring & Horrocks, 2001).

As we saw in the revolution that accompanied P2P business, digital communications are also substantially changing political processes so that we can clearly see the peer production of political content. In both cases, what has really increased is the participation of citizens when compared with the structures and systems of traditional organizations.

This is why Benkler and Nissenbaum defend the whole peer production system and want to present it as a change in the anthropological dimension of society at all levels. For these authors: "the emergence of peer production offers an opportunity for more people to engage in practices that permit them to exhibit and experience virtuous behaviour" (2006, p. 394). It is an anthropological change that leads citizens to develop a range of civic virtues such as independence and autonomy and which encourages creativity, social engagement, altruism and cooperation, among others (Benkler, 2006).

In short, disintermediation, from this perspective, is generating a new communicative scenario by increasing, among other things, citizens' ability to express their demands, to coordinate on issues that are of collective interest and to respond quickly to social and political processes that interest them. Ultimately, at least in theory, the effect of technological development that we are calling "disintermediation" will generate a scenario that should favour more diverse public opinion. However, as we said in the introduction, using the words of Papacharissi (2002), there is a new public space for politically oriented conversation; whether this public space becomes an arena for exchanges between agents of public opinion or not does not depend on the technology itself. We shall leave this question for later and focus on how we can apply disintermediation to the specific practical actions by citizens and organizations, advancing in small cautious steps in order to proceed on solid ground to the greatest extent possible.

References

Altheide, D. L., & Snow, R. P. (1979). *Media logic*. Beverly Hills, CA: Sage.

Bell, D. (1973). Technology, nature and society: The vicissitudes of three world views and the confusion of realms (pp. 385–404). *The American Scholar*.

Bell, D. (1976). The coming of the post-industrial society. *The Educational Forum, 40*(4), 574–579.

Benkler, Y. (2006). *The wealth of networks: How social production transforms markets and freedom*. New Haven, CT: Yale University Press.

Bennett, W. L., & Segerberg, A. (2013). *The logic of connective action: Digital media and the personalization of contentious politics*. New York, NY: Cambridge University Press.

Castells, M. (1977). *The urban question: A marxist approach*. Cambridge, UK: MIT Press.

Castells, M. (1996). *The information age: Economy, society and culture, Volume I: The rise of the network society*. Cambridge, UK: Blackwell.

Castells, M. (2012). *Networks of outrage and hope: Social movements in the Internet age*. Cambridge, UK: Polity Press.

Entman, R. M. (1993). Framing: Toward clarification of a fractured paradigm. *Journal of Communication, 43*(4), 51–58.

Gibson, R., & Ward, S. (2009). Parties in the digital age, a review article. *Representation, 45*(1), 87–100.

Hallin, D. C., & Mancini, P. (2004). *Comparing media systems: Three models of media and politics.* Cambridge, UK: Cambridge University Press.

Hjarvard, S., Mortensen, M., & Eskjær, M. (2015). Three dynamics of mediatized conflicts. In M. Eskjær, S. Hjarvard, & M. Mortensen (Eds.), *The dynamics of mediatized conflicts* (pp. 1–27). New York, NY: Peter Lang.

Jenkins, H. (2006). *Convergence Culture: Where Old and New Media Collide.* New York: New York University Press.

Jiménez, G. (2014). El proceso de desintermediación comunicativa. *Revista Internacional del Mundo Económico y del Derecho, 7,* 69–91.

Klinger, U., & Svensson, J. (2015). The emergence of network media logic in political communication: A theoretical approach. *New Media & Society, 17*(8), 1241–1257.

Knorr Cetina, K. (1998). Sozialität mit Objekten. Soziale Beziehungen in posttraditionalen Wissensgesellschaften. *Technik und Sozialtheorie, 42,* 83–120.

Lasswell, H. D. (1927). *Propaganda technique in the world war.* Cambridge, UK: MIT Press.

López-Escobar, E., Llamas, J. P., McCombs, M., & Lennon, F. R. (1998). Two levels of agenda setting among advertising and news in the 1995 Spanish elections. *Political Communication, 15*(2), 225–238.

Lutz, C., Hoffmann, C. P., & Meckel, M. (2014). Beyond just politics: A systematic literature review of online participation. *First Monday, 19*(7), 1–36.

Mansbridge, J. (2003). Rethinking representation. *American Political Science Review, 97*(4), 515–528.

Mansbridge, J. (2011). Clarifying the concept of representation. *American Political Science Review, 105*(3), 621–630.

McCombs, M. E., & Shaw, D. L. (1972). The agenda-setting function of mass media. *Public Opinion Quarterly, 36*(2), 176–187.

Meyrowitz, J. (1993). Images of the media: Hidden ferment—And harmony—In the field. *Journal of Communication, 43*(3), 55–66.

Olson, M. (1978). *The logic of collective action: Public goods and the theory of groups.* Cambridge, MA: Harvard University Press.

Papacharissi, Z. (2002). The virtual sphere: The internet as a public sphere. *New Media & Society, 4*(1), 9–27.

Pitkin, H. F. (1967). *The concept of representation.* Berkeley: University of California Press.

Rehfeld, A. (2011). The concepts of representation. *American Political Science Review, 105*(3), 631–641.

Sampedro, V. (2014). *El cuarto poder en red. Por un periodismo (de código libre) libre.* Madrid, Spain: Icaria.

Sartori, G. (2005). *Parties and party systems: A framework for analysis.* Colchester, UK: ECPR press.

Schradie, J. (2011). The digital production gap: The digital divide and Web 2.0 collide. *Poetics, 39*(2), 145–168.

Stehr, N. (1994). *Knowledge societies.* Thousand Oaks, CA: Sage.

Strömbäck, J. (2008). Four phases of mediatization: An analysis of the mediatization of politics. *The International Journal of Press/Politics, 13*(3), 228–246.

Tuchman, G. (1978). *Making news: A study of construction of reality.* New York: Free Press.

Wring, D., & Horrocks, I. (2001). Virtual hype? New media and the transformation of political parties and party systems. In B. Axford & R. Huggins (Eds.), *New media and politics* (pp. 191–209). London, UK: Sage.

Zabludovsky, G. (2013). El concepto de individualización en la sociología clásica y contemporánea. *Política y cultura, 39*, 229–248.

CHAPTER 4

The Political Potential of Social Networks

In Chapter 2 we have pieced together a rough outline that describes a process of social change from a stage of representation and communication in which public opinion is created by permeable (consumers of information) and observable agents to a society with the resources and capacity for critical organization through social media.

In order to advance in our explanation of the disintermediation process, we propose to refer readers to a more specific territory. This territory would be defined by actions carried out by people and organizations in a technological context that is dominated by disintermediation.

In this sense, we can identify three disintermediation processes that affect the actions of the social agents and, in general, their participation in public opinion. These three processes are the disintermediation of the agents, of the messages and of the spaces for communication. These "specific disintermediations" will enable us to define how digital technologies generate the transformation in communications that we outlined above.

However, we must bear in mind that these three processes correspond to the potential and idealized dimension of disintermediation. With this, we mean that both the disintermediation of agents and the disintermediation of spaces and messages are the aspects that, in ideal terms, could help to generate a more open, inclusive and critical public space. Naturally, between the potential of this process and practical reality, there are usually substantive differences. Throughout this chapter, we will try, not only to

© The Author(s) 2019   37
J. M. Robles-Morales and A. M. Córdoba-Hernández,
Digital Political Participation, Social Networks and Big Data,
https://doi.org/10.1007/978-3-030-27757-4_4

explain the potentialities of disintermediation, but also to outline some of the most controversial issues that will be taken up throughout the second part of this book and, especially, during the conclusions.

4.1   THE DISINTERMEDIATION OF THE AGENTS

Not long ago, a friend was describing some of her earliest experiences with the public administration, no longer as a bystander but the subject of the process, specifically to obtain a passport. She was due to travel abroad a few weeks later and was tremendously excited about it. All her excitement drained away, however, when she found herself at the end of a long queue of people waiting to obtain their passport from a single window at the end of a long, gloomy corridor. When it was her turn, a man should she could barely see through a cloud of smoke (people were still allowed to smoke at work back then) asked her for a series of documents that she did not have, so she had to return the next day and queue up all over again.

In this same period, she was taking her first steps in the world of activism and she recalled how much work she and her young friends had to do to convince her classmates, much more interested in football and dating, that they needed to go on strike to avoid losing the access they enjoyed to one of the city's theatres. She also recalled how many times she had called her public representative at that time by telephone to explain what she was campaigning for and for him to support her cause. Naturally, he never returned her calls and she was left with her first taste of the sensation of distance existing between politicians and herself. This sensation would accompany her for the rest of her life.

Finally, despite obtaining the support of some of the teachers, our friend told us of the difficulties she faced in finding resources that would allow her to buy the few things she needed to be able to make banners and pamphlets. Once she had them, she had to recruit one of the smarter kids from her class to make an attractive design and to write the messages for the pamphlet that would be sure to change the mind of all the youngsters in the world.

Although we could not see it at the time, these episodes were closely connected. They were our friends' first experiences as a citizen of events that many more of us have gone through. Given that she was too young to pay tax, her contacts with the public administration were limited to driving licences, passports and other similar processes. On the other hand, as a citizen, she was now in a position to make her demands public and make demands through collective action. This step was not an easy one to make. The institutions that she was addressing (we were all addressing) were inflexible and the actions to take were very costly. To carry out any process with the administration requires a lot of free time within the working hours of 8 am to 2 pm. On top of this, it is not easy to find the information about the documents we need to submit, which makes the process inefficient (basically, there is always one more document missing). In this way, every official appointment means losing at least a morning and, of course, several hours of class.

When it comes to activism, convincing young people who, like us, were more interested in changing things for the better than playing football, was difficult and frustrating. So was getting the attention of those who had power or getting enough resources to force them to pay attention (through banners and pamphlets). There were obstacles everywhere that reinforced our strength of character as citizens.

Even so, all those present at the meeting where our friend told her story agreed that young people would not have this same experience today. The Internet has practically taken those obstacles out of the picture. What we now refer to as e-government has eliminated many of the barriers that we saw in the story about the passport. And cyberpolitics is significantly lowering the cost of producing content that makes collective action much easier to arrange.

The term e-government, or digital or electronic government, includes all the measures intended to modernize governmental bureaucracy and activity by taking advantage of the Internet as a platform for connectivity. Cyberpolitics refers to the use of the Internet for political participation. Experts draw on two basic concepts in the social sciences to define these Internet-based forms of participation; conventional political participation (Verba, Nie, & Kim, 1978) and unconventional political participation (Barnes, Kaase, & Allerbeck, 1979). We shall use *conventional digital political participation* to refer to political actions that derive from government activity in the framework of representative democracy. *Unconventional digital political participation* will refer to activism and other forms of participation (Robles, De Marco, & Antino, 2013).

The Internet has widened the range of opportunities for citizens to participate through both e-government and cyberpolitics. These opportunities are essentially the outcome of lower participation costs, so that there are many more actions available to citizens through cyberpolitics, and in the case of e-government the types of interaction are more flexible and efficient.

Today, social media makes it easier than ever for any citizen to get in touch with other people who, like them, want to change the world in a particular way. Platforms such as Twitter or Facebook also make it possible for users to bypass the financial restrictions that young people faced in the analogue age when we wanted to make banners and pamphlets, hoping to make ourselves heard. In today's world, making an issue public only requires you to create an attractive hashtag that people notice and to wait for it to come to the attention of others. At the same time, thanks to the Internet, political parties and representatives are potentially more exposed to direct messages from the electorate and it may no longer be necessary to make dozens of phone calls, like we did in the nineties, in order to get a response.

The point is, to what extent have the institutions that we have to deal with to handle our public affairs and the organizations that represent us adapted to this new digital scenario and, by extension, this new way of engaging with ordinary citizens? On the other hand, to what extent does this process really mean an improvement in the dialogue with political parties, public representatives and the administration?

As we noted above, e-government means improved communications possibilities in several directions: between citizens and the authorities, between these authorities and private sector organizations, and between the government and civil servants (Jeong, 2007). It also influences, at least in theory, initiatives that aim to develop and expand democracy in the digital environment (Koh & Prybutok, 2003).

There are already a number of indexes and indicators in this sense that measure the extent to which developed countries are achieving these goals. One of the most commonly referred to is the United Nations' E-government Readiness Index (EGDI). This Index is the result of three sub-indexes: for online services, technological infrastructures and human capital. Through these we can determine the extent to which the countries studied possess a policy for digitalizing their activities and services or not; whether citizens have the tools and resources for joining the digital society, and finally, if there are citizens able to take part in this transformation.

According to the latest ranking, published in 2016, these are the 25 countries with the best electronic government. As you can see, they are highly developed countries and most of them are from the developed world, which supports the perception that there are asymmetries and inequalities that affect the political developments that we are discussing (Table 4.1).

Table 4.1 List of the top 25 countries in the world on the EGDI index 2016

Country	Region	EGDI 2016	Ranking 2016	Ranking 2014	Ranking 2012
United Kingdom	Europe	0.9193	1	8	3
Australia	Oceania	0.9143	2	2	12
South Korea	Asia	0.8915	3	1	1
Singapore	Asia	0.8828	4	3	10
Finland	Europe	0.8817	5	8	3
Sweden	Europe	0.8704	6	14	7
Holland	Europe	0.8659	7	5	2
New Zealand	Oceania	0.8653	8	9	13
Denmark	Europe	0.8510	9	16	4
France	Europe	0.8456	10	4	6
Japan	Asia	0.8440	11	6	18
United States	America	0.8420	12	7	5
Estonia	Europe	0.8334	13	15	20
Canada	America	0.8285	14	11	11
Germany	Europe	0.8210	15	21	17
Austria	Europe	0.8208	16	20	21
Spain	Europe	0.8135	17	12	23
Norway	Europe	0.8117	18	13	8
Belgium	Europe	0.7874	19	25	24
Israel	Asia	0.7806	20	17	16
Slovenia	Europe	0.7769	21	41	25
Italy	Europe	0.7764	22	23	32
Lithuania	Europe	0.7747	23	29	29
Bahrain	Asia	0.7734	24	18	36
Luxembourg	Europe	0.7705	25	24	19

Source E-government survey 2016. E-government for the Future we want, United Nations 2016

The public policies for the development of the Information Society, both within the European Union and many of the most influential countries in the world, are giving priority to systems for citizens and the government to interact without intermediaries and in more direct and flexible ways. In other words, a relationship in which information is more freely available (open data), procedures are standardized and citizens more autonomous when it comes to handling their affairs with the public administration.

However, this process is done at the expense of the fact that important layers of the population are cyber-excluded. We know from various studies that digital inequalities generate great difficulties for certain social groups to be incorporated into this type of advanced Internet use (Robles, De

Marco, & Antino, 2014). For example, people with low levels of education, the elderly or people with less economic resources, despite having joined the use of the Internet, do not have enough motivation and digital skills to make use of e-administration (Robles, De Marco, & Molina, 2010). Despite what may seem to many of us, online administrative procedures require skills and abilities that are not always immediate and intuitive. This being so, and although there is enough empirical evidence, European governments and other regions of the world, propose the closing of the analogue window (that window behind which the smoker used to be). The closing of the analogue window means, today and in practical terms, the digital exclusion of many citizens and a severe limitation of their rights. The development of e-government and e-administration, even with its undeniable potential advantages and its disintermediating nature, is like a train that not all citizens got on the departure station (the station would be using the Internet) but that, even worse, as its speed increases, it loses more and more people who do not have a first-class ticket.

Let's return to the concept of cyberpolitics that, as we saw earlier, should be understood as categories of digital political participation that can be either conventional or unconventional. In the first studies on the relationship between politics and the Internet, the practices noted under this heading were seen as clear examples of the magnificent opportunities offered by the medium. It was considered that the Internet favoured the links between representatives and those they represent and would have a positive effect on democratic growth (Hague & Loader, 1999). In later studies, however, other authors were less optimistic, claiming that the Internet had less of an effect (Boulianne, 2015), a debate in which academics have not come to an agreement on, in which some argue that only the traditional forms of action that require greater personal commitment on the part of citizens should be considered political, while others hold that political actions are evolving and finding new forms of expression.

We have become increasingly accustomed to different manifestations of cyberpolitics in recent years. For example, electronic voting, which can mean both the method of voting or remote voting on the Internet and the benefits provided by electronic media in the processes of counting and organizing the electoral process (Gálvez Muñoz, 2009).

We can also see the widespread use of websites by the main state institutions, such as the government, ministries, city councils and other public bodies. Specialists have pointed out in this field that there are different strategies for incorporating the Internet into political party work, such as the use of websites (Dader, 2009), the use of digital social media in election campaigns (Abejón, Sastre, & Linares, 2012), the personal blogs for politicians and a host of other strategies for spreading electoral messages over the Internet (Anduiza, 2009; Sampedro, López, & Muñoz, 2012).

However, the incorporation of this by political representatives comes at the cost of a low level of democratic innovation and limited capacity to include their supporters' proposals in electoral manifestos or to take positions on specific issues. This is why many researchers emphasize that the engagement of many public institutions in today's digital environment draws on the standards of Web 1.0, which is geared to work in one direction only, and not 2.0, where a true approximation could be expected between representatives and the represented, thanks to the digital tools that enable communications to flow horizontally.

Recent studies show positions which go some way towards modifying this diagnosis of political parties using the Web 1.0 approach in countries like the United States, Spain, Germany and Italy. For Bennett, Segerberg, and Knüpfer (2018) the dialectic of the parties on the right and left has been disturbed by the emergence of "connective parties" that incorporate the potential of the Web 2.0 and are characterized by opening their central organizing functions to citizens and supporters acting through digital platforms.

These new collectives want to distance themselves from the organizational strategies of the right-wing, which generally favours hierarchies and strong leadership based on shared values, and those of the left that seek to act as standard-bearers for diversity with political practices that favour inclusion and deliberation. It is in this scenario that digital technologies are seen as a structure of opportunity. Some traditional political parties, but mainly the new left-wing parties that are appearing in Europe have been proposed as examples of this connective process. For example, the Spanish political party *Podemos*, which was formed in 2014, used an online platform, *appgree* to channel the debates and proposals made by all the citizens who wished to take part in establishing the bases of its electoral programme in 2016.

It is not only the strategies of connective parties that are opening up the arena of conventional digital political participation. There are other groups, businesses, non-profit organizations and applications such as *Civitana*, *Change.org* or *Appgree*, mentioned above, that are moving in the same direction and offer digital tools and services to make popular consultations or for collecting signatures for petitions on certain political or social issues. This is clearly an important step that has the potential to commercialize the political attitudes, values and activities of citizens who are playing an increasingly important role.

However, several recent studies have shown that, although, in the first instance, these political parties emerged with a desire for horizontalization and inclusion, their actual practices are far from reflecting a real openness (De Marco, Robles, & Gómez, 2019). There are two basic problems in the inclusive practices of political parties that have arisen in the digital context. The first is the availability to incorporate the opinions that citizens express online to the electoral programmes of these parties or to transform the citizen's opinions into the base for real public policies. That is, political parties, even those called connective, rarely take the feedback that citizens offer them through social networks as a basis for their decisions. This circumstance generates an effect of discouragement among citizens who perceive this closure as another example of the online version of traditional party strategies.

In this same line, far from interacting, debating and responding to the inquiries of citizens through social networks such as Twitter, are limited to disseminating information (De Marco et al., 2019). This same study shows how, the public representatives are much more inclined to respond to those interpellations made by journalists or other politicians than those made by citizens. This, again, supposes a bet to favour the traditional mediators to the detriment of the new citizen mediators.

That being the case, the question is to what extent the reorganization of the political parties that emerged in the digital era is nothing more than a strategy for gathering followers in a context in which citizens have the sensation of being more listened to. Our impression is that, despite a promising scenario, in which agents favour disintermediation strategies, real practice generates some important imbalances that severely limit this potential. The "traditionalization" of the strategies of the parties that emerged in the Internet era indicate something that was already reflected after the disintermediation in the economic sphere. That is, traditional agents tend to absorb the synergies of disintermediation to make them their own and favourable to their interests. This is to adopt an image of horizontality and openness to the digital debate to respond to a social demand without implying a real and deep compromise with the new political challenges.

4.2   The Disintermediation of the Messages

We can refer to the first core aspect of the disintermediation of the messages as information autonomy. According to this idea, the ease of accessing information and Web tools means that users can go directly to the original sources and leapfrog the interpreting filters. This autonomy that people acquire influences the whole process of political communication because it may encourage direct contact with other actors and multiplies the possibilities for contacts and relations. It was unthinkable in the past for citizens to have direct access to the full, original speeches made by public figures. Nor could they address a politician directly or openly express their opinions on the public stage without having to write a letter or look for a call-in space on a radio station.

According to Batle and Cardenal (2006), information autonomy has led to a drastic decline in the dependence on traditional channels of information, to the point where intermediaries are overlooked or relegated into the background. On this way, other authors highlight how the surfeit of data and interactivity on the Internet has made it possible for those who used to consume news and political messages in a purely passive way to seek other sources of news and challenge official discourse (Loader & Mercea, 2011). However, this is not about giving vent to optimism. This behaviour, itself, has nothing new. Every wave of social change has attempted to use the technologies available with a dual intention, to alter the status quo in public opinion and to share political and economic power.

From this point of view, it is understood what Deuze argued, we find ourselves immersed in the digital culture of the twenty-first century after passing through the electronic culture of the twentieth century and the print culture of the nineteenth century (2006). The proliferation and saturation of media on all kinds of screens forces us to reconstruct information using three main processes: (a) *Participation*, in the sense that as consumers we become active agents in the construction of meaning; (b) *Remediation*, because the user blends together information acquired from various sources and adopts new meanings to understand the real world and (c) *Bricolage*, in which the user carefully builds new personal versions of this reality to share with others.

We should not ignore that fact that these options for handling information work both ways, in the sense that public actors and organizations also have the option of broadcasting their actions and discourse beyond their local boundaries, seeking support and recognition on a larger scale

(Lago & Marotias, 2007). If this dynamic is achieved, it would offer even greater autonomy of action, without dependence on central nodes of organization, something that may allow increasing the opportunities to spread other points of view, design one's own strategies or to replicate those of other users and groups (Sánchez & Magallón, 2016).

The disintermediation of the messages can also be understood from the same production of the contents that are transmitted. Largely, digital media has enabled an important change in communicative processes in which, it could be said that, the passive role of traditional audiences as the receivers of messages, lead a change of roles in which the same receiver now often plays the role of message issuer, without giving up their position as consumers of media. This is the situation that has been defined in the "prosumer" concept, proactive consumers who do not act for profit and whose only motivation is to share information in the digital sphere, where they can generate as much or even more information that the traditional organizations (Islas & Arribas, 2010).

For Contreras (2013), All the resources for generating digital content can be seen as the beginning of a new culture that bases its success on user collaboration, leaving behind the old three-phase model: production, channel and consumption, to create space for processes of popular appropriation, with collaborative technological resources. As Scolari (2008) argued, we are looking at a rupturing of traditional categories and a movement from communication consumption to production in which the user finally becomes part of the content. This is why he says, paraphrasing McLuhan, that in the new forms of digital communication the user is the message. A message that acquires more value and importance the more it is shared.

The opportunities that prosumers have to spread their ideas on different platforms and social networks mean that we can talk about horizontality as it was foreseen in the earliest days of the Internet. Without entering the debate on the degree to which this horizontality exists today, it is true that authors as Benítez (2013), affirm that many of the message transmission channels are not necessarily regulated by relations of power but come from other actors and how this results in the disintermediation of the messages that we have been discussing.

What has been stated up until this epigraph, can be understood from the structure itself of the Internet that is conceived as a "many-to-many" model that renders obsolete the classic "one-to-many" paradigm on which the old theories of mass communication broadcasting were based. Thanks to its *reticular*, in Scolari terms-or *multilateral* according to Cotarelo (2010), the Internet nature has become an open space in which any individual can participate, despite having a strongly critical opinion. It is a place in which antagonistic values can co-exist, although they may share the same cultural basis. The fact that any person can criticize the public statements of a politician on the wall of the same politician's personal account, express their opposition or expand the debate by adding new content, reveals that, at least, the system for the creation, control and transmission of messages is changing.

Throughout this section, the reader will have noted the on-going relationship we have been establishing between citizen-users and the messages they send. We cannot consider the true disintermediation of these messages without understanding what their authors do. These prosumers are at the heart of the communication process. They may not know each other, can exist in the hundreds, thousands or millions, and be spread out across the world, but they react in a similar way to different stimuli and find themselves in a position to influence the traditional system of political action from their platforms, networks and screens.

It is interesting at this point to refer back to the concept of *connective action* (Bennett & Segerberg, 2013), that we mentioned earlier. The communicative flows we are seeing exhibit a strong tendency for the digital citizen to personalize political content through social media, which does not require great commitment or firm identification with the cause or campaign that is supported, often no more than clicking on a button, sharing a link or distributing something of interest to extend the reach of a point or to raise its volume when it seems to be absent from the media agenda.

If this is so, what is the basis for success in a global campaign or mobilization on the Internet? According to Bennett and Segerberg (2013), in how easily a message can be adapted and personalized in different contexts. To illustrate with an example, we can see what happened with the slogan of *Occupy Wall Street*, in 2011, the famous "*We are the 99%*", which is seen as the most effective political slogan in American politics since "*Yes, We can*" in Barack Obama's 2008 presidential campaign.

The author of the slogan, Chris, a New York journalist working on a local paper used it for the first time in August 2011 on his *Tumblr* blog to encourage visitors to publish a photo with a message explaining how the financial crisis had affected them. The response was immediate and in less than six months more than 3000 users had added their photos with panels, pages and notice boards in which they explain their situation. After all, who would not feel like a part of the 99% of the population as opposed to the

1% of the financial elite who hold the wealth of the country? Later, in the Occupy Movement's tent city in Zucotti Park, the slogan would spread rapidly on social media and from there it would make the leap to the main media outlets and the political sphere, going on to be named as the "phrase of the year" by Yale University and quoted explicitly by President Obama in a speech on social inequality (Peinado, 2011).

As we can see in this example from New York, the adaptability of content calls for flexible frameworks that can be shaped, expanded and modified by users to suit their personal lifestyles. After all, these will make up the agenda of their individual concerns and it is from here that they will find points of contact with other citizens (Sánchez & Magallón, 2016).

In these political processes, the power of a social network derives from its potential to allow users to take control of content and extending it to other social spaces, the engagement of individuals with similar interests for whom the message will be particularly apt (Dahlgren & Álvares, 2013; Villanueva-Mansilla, 2016). Unlike the political sphere, which understands *engagement* to mean direct participation in the affairs of political life, on digital social media it means the capacity for generating reactions on the part of users.

In the digital environment, personalized communication consists, as Bennett and Segerberg (2011) and Benítez (2013) said on of providing people with greater opportunities to define subjects on their own terms and to interact with others (Benítez, 2013; Bennett & Segerberg, 2011). It is a connection through shared activities, but always guided by self-expression, and it is made possible by technology. With this, according to Contreras (2013), people are participating in connective activity without sacrificing individual identity, whether publishing content in isolation in closed spaces or sharing the posts of other users in open spaces.

To end this section, we would like to add a final thought on the content, because it is the fact that traditional organizations no longer have such a leading role which has turned the messages themselves into the place where users come into contact with each other. Citizens often find that they have more in common with others, not due to the traditional areas of identification, such as family, employment, religion, nation or politics, but because they find themselves sharing similar content with them with.

These digital connections have been a constant target of criticism and mistrust on the part of academics who feel that these actions are based on very weak connections, sometimes preferences or hobbies, or other times on shared emotions or needs such as indignation or anger (Bernete, 2013).

It is true that Internet entails a degree of ambivalence in terms of identity, because while these digital communication channels offer freedp, and the expansion of opinions and identities that we can embrace, they do so at the expense of elements as important as traditional collective action, responsibility and personal commitment (Sampedro, 2006).

In this sense, we can say that those who identify as "digital citizens" show a preference for "liquid identities", for intermittent, occasional militancy, flexible and volatile commitments that do not require one to bind oneself permanently to any organization. This is why authors such as Bernete (2013) recommend using the term *identifications* rather than identities to emphasize the occasional and elective nature of movements of uncertain duration, in which the people involved retain greater autonomy.

As Kuklinski explained in an interview, people need not connect to structured political institutions, but only enter into digital contact to arrange something and then disconnect when the goal has been achieved. This is why there are so many citizen movements, or *smart-mobs*, and there is a view of cities as interfaces where people connect with each other digitally, or at least analogically but guided by a digital approach, without vertical structures and pre-established hierarchies (Fraticelli, 2013).

The emergence of these sporadic identifications with specific causes or issues starts with the selective consumption of information and messages. Several recent studies have shown that citizens tend to consume information that is aligned with their tastes and opinion and that this selective exposure reinforces their beliefs and avoids sources that challenge them (Bennett & Iyengar, 2010; Sánchez & Magallón, 2016).

In the case of social networks such as Facebook, Twitter or Instagram, there has been a general fragmentation and polarization of political communities that has encouraged the creation of a "spiral of selective attention" (Neuman, Bimber, & Hindman, 2011) that backs up preconceived ideas and which can be used by political organizations and actors to determine discourse, confirm group values and promote their ideals (Dalhgren, 2005).

Normally, individuals become loyal to certain social networks or websites because of the sense of belonging to the same ideological spirit or the same lifestyle. One of the characteristics of online communication consists precisely of user allegiance to content because of the loyalty they feel towards specific channels and their new forms of relationships with media content (Contreras, 2013; Jenkins, 2006).

In this same vein, the fact that we talk of sporadic identification and not people's firm identities, has a direct effect on the forms of militancy in political causes. We could say that the traditional idea of membership has expanded thanks to the Internet, because digital citizens demand different ways of taking part that are less exclusive, more flexible and which offer diverse paths and degrees of involvement (Van Laer & Van Aelst, 2010). The key to the political participation of this new citizen raises questions about militancy itself, because this type of identification with political causes rejects the demands of unvarying, uncritical commitment (Sánchez & Magallón, 2016).

The collaboration that is offered to a specific campaign or organization owes more to the activation of politically informed and motivated citizens whose militancy is on standby, staying at the margins in a latent condition that is activated by political opportunity and motivating circumstances (Amnå & Ekman, 2014). Activity occurs when it is necessary and always requires that the grounds and context are in place to ensure that it will be effective. We could describe it as "pre-political" behaviour, by citizens who debate politics, consume news and talk of social issues and are inclined towards participation.

As Subirats claims, it is no longer necessary to be a "militant". A digital citizen can be part of different projects at the same time, in a type of "political promiscuity" that easily changes from one cause to another and collaborates occasionally or in response to particular proposals, even when no commitment is specified. Therefore "the way that young people engage with political spaces today is closely related with these "liquid" forms of commitment and the decline in the importance of stable political identities" (2015, p. 128).

4.3   THE DISINTERMEDIATION OF POLITICAL SPACES

The first person to use the term cyberspace was the American author William Gibson in his science fiction novel *Neuromancer* (1984). At the time, the author wanted to describe the burgeoning digital network in terms of a spatial metaphor, a figure that was so successful that it entered mainstream language to describe interactions on the Internet as a type of alternative space or virtual paradise.

The debate about cyberspace and its scope had barely started. Time has shown that the political impact of Internet is partially dependent on the modification of the concept of space and the resulting transformation of classic binary formulas like public–private or local–global. In the dynamics of the digital environment, the user is constantly moving from one place to another, from private to public, from real to virtual, so that we can agree with Agra (2012) that the Internet merges the duality of the physical and virtual, so that time and space lose their linear nature.

There has been a lot of theorizing about this aspect of the Internet. Bauman (2002) considers it to be a crucial attribute that distinguishes the progress of the history of modernity and from which the other characteristics of this period emerge. For this author, modernity begins when space and time are separated from everyday experience and are treated as independent categories.

First of all, he alludes to the sudden irrelevance of space, disguised as the annihilation of time. Today we can cover space literally "in a fraction of the time" and the difference between near and far has disappeared. "Space no longer limits either the action or its effects and has little or no importance at all. It has lost its 'strategic value' as military experts would say" (Bauman, 2002, p. 126).

On the other hand, because the Internet can only have a transnational, global character, it seems that space has also been deprived of its physical aspect there. In digital media, it is technology that defines the limits of the space, offering a level of control over how the interactions take place that is far greater than in the real world. Problems can be programmed or "coded" just as they can be "decoded" when we want them to be.

As the father of the Internet, Berners-Lee (2000) said, the web was never conceived as a physical "thing" that existed in any specific place, but as a "space" in which information could exist.

Because of the above, and the way in which it is coordinated, digital citizens inevitably enter a certain *deterritorialization* (Deleuze & Guattari, 1983), understood as the loosening of the natural link between culture and geographic and social territories. The Internet generates a mixed, hybrid culture that encourages socialization based more on alignment than membership, and which opens more doors (Gómez, 2004; Lemos, 2002).

In the early days of the Internet, Castells (1996) already looked into this aspect when he affirmed that the Net had its own geography, consisting of nodes and networks that processed information flows and were controlled from certain points. However, while he acknowledged that the interaction of these flows generated a new type of space, he exercised caution in claiming that this was a redefinition of distances, but not the elimination of geography.

Therefore, and even though "deterritorialization" supposes a potential for the dissemination of information, the capture of followers for a concrete political cause or access to knowledge, the practice implies a superficial approach to the political and social processes that based on often spurious similarities and superficial knowledge of events. Knowledge that, in many cases, is based on biased and partial information to which the truth value is given based on its spectacularism or immediacy. In the same way, deterritorialization tends to treat problems without considering the specific and local issue. That is as if all the issues could be measured by the same pattern that is generally the Western pattern. In short, we refer to the fact that deterritorialization, despite its potential, generates a false image of closeness between the supporters and online activist and the real problems that affect people, as well as a false universalization of solutions.

According with Benítez (2013), this feature of Internet's reshaping of space and time can have a positive effect on digital participation, should this cross-boundary culture result in more transparent information flows for decision making. The discourses transmitted through platforms and social media are not necessarily mediated by power relations but can emerge from other actors through horizontal structures (Benítez, 2013).

One example will show this. We shall take a look later at the case of the hashtag #BringBackOurGirls, which politicians, celebrities, the media, citizens and organizations around the world used on social media to express rejection and concern for the kidnapping of 276 girls by Boko Haram in Nigeria in 2014. This case will serve us as a reference point to see one of the ways in which the digital environment is permitting the disintermediation of places through what Keck and Sikkink (1999) called the Boomerang Effect.

The Boomerang Effect occurs when local social actors seek to go beyond the limits of their own states and achieve international support with which to press their authorities. Just like a boomerang that rises after being thrown to return to its point of origin, a cause that becomes global and succeeds in achieving the adhesion of activists in developed countries makes it possible for local actors to "amplify" their demands with this echo from without, returning to the national political arena with greater force.

If we focus now on political participation, the Internet has also managed to create and recreate "spaces" that enable debate and deliberation on topics of common interest (Karakaya, 2005). We can say that Internet has expanded the topics in the public arena, but also modified the way in which these are appropriated, using technology to enable lines of individual participation to emerge, either gradually or rapidly, by going viral.

Of all these spaces, the social networks, where control of planning and content is surrendered, have assumed a role as natural places for political activity. For Papacharissi (2010), platforms like *Facebook* or *Twitter*, represent functional strategic support for working in collaboration and in contact with other users, despite the lack of consensus among them at times. It is the technological tools uniting us that are important, although they do not make us culturally or ideologically the same. The construction of these places for free participation opens up paths to knowledge and facilitates the acceptance of differences, because the space created by digital media reduces cultural divisions through creative work under participative dynamics.

According with Galindo (2013), *Facebook* is the perfect niche for a new emergent culture, made of scraps, bound together in ephemeral, constant motion and barely brushing the surface. This complexity is permanently in flux, never seems to settle down and leaves no sediment and acquires no depth. It is centred on the subjective user experience more than shared interests and knowledge.

As regards the time dimension, Bauman considered that time has ceased to be a value that can maximize the return on the value of space. On the contrary, we are facing "instantaneousness", rapid movement in the briefest of moments that reveals the absence of time as a factor in an event.

> Power can move at the speed of an electronic signal, so the time required for the movement of its essential components has been reduced to instantaneousness. In practice, power has become truly extraterritorial and is no longer bound, or even held up, by the resistance of space. (Bauman, 2002, p. 16)

This new conception of time is what has come to be called "real time", in which everything is interconnected (Barabási, 2002). The key aspect of this study is to ask to what extent real-time communication has changed forms of action and enabled different ways of acting politically. Cyberspace is the reality of the moment which is created in the form of discourse. It is immediate, and the immediate is by definition the here and now.

Rheingold (2002), in his book *Smart Mobs*, shows how the ability to make decisions on the fly makes Internet users wary of dividing their lives into temporal categories, leading him to ask: Has physical space become dissociated from the definition of "presence"? Does the concept of a social network that reaches beyond a single specific space apply today?

Some other authors are even more critical with the question of the speed and instantaneity of time in the network. Thus, for example, Virilio (1986) alerts us to the negative impact that the speed of technology has on the field of politics. Virilio's great legacy is his idea of accidental time. Until the massive irruption of digital technologies and global and viral media, there were two times. On the one hand, we had long-term time, that is, the history that cares about the processes and structure of societies. On the other hand, we have the time of events. That is, those facts that we can call historical because of their social, political or economic importance and scope.

However, the accidental time, says Virilio, is an instant that does not participate either in the past (it is decontextualized) or in the future (it will disappear in a short time). A time, in which Snapchat users live, and that is governed by the accident rate of the most immediate present. Speed is the fundamental movement of accidental time and digital technologies are the weapon that makes this movement effective.

The corollary of Virilio's argument is that, by understanding our lives as accidental time, we cede the power of our existence to machines. The speed of accidental time is not the speed of reflection and political debate. It is a speed that only corresponds to the machine. The political messages that we post on Twitter and that we briefly share with the digital community, do not reflect our political concerns and demands, but the moments of our lives and fleeting inspiration.

References

Abejón, P., Sastre, A., & Linares, V. (2012). Facebook y Twitter en Campañas Electorales en España. *Anuario electrónico de estudios en Comunicación Social, 5*(1), 129–159.

Agra, S. (2012). El sujeto hipertextual: la desterritorialización en la comunicación mediada por ordenador. In *Actas de The 10th World Congress of the International Association for Semiotic Studies (IASS/AIS)* (pp. 1479–1487). Universidade da Coruña.

Amnå, E., & Ekman, J. (2014). Standby citizens: Diverse faces of political passivity. *European Political Science Review, 6*(2), 261–281.

Anduiza, E. (2009). Internet, campañas electorales y ciudadanos: el estado de la Cuestión. *Quaderns del Cac, 33*, 5–12.

Barabási, A. (2002). *Linked: The new science of networks.* Cambridge, UK: Perseus Publishing.

Barnes, S. H., Kaase, M., & Allerbeck, K. R. (1979). *Political action: Mass participation in five Western democracies.* Beverly Hills, CA: Sage.

Batle, A., & Cardenal, A. S. (2006). La utopía virtual: Una crítica al ciberoptimismo desde la teoría de la elección racional. *Revista de Internet, Derecho y Política, 3,* 1–12.

Bauman, Z. (2002). *Modernidad líquida.* Buenos Aires, Argentina: Fondo de Cultura Económica de Argentina.

Benítez, L. (2013). La dimensión transnacional de la ciudadanía digital. En F. Sierra (Ed.), *Ciudadanía, tecnología y cultura: nodos conceptuales para pensar la nueva mediación digital* (pp. 79–118). Barcelona, Spain: Gedisa.

Bennett, W., & Iyengar, S. (2010). A new era of minimal effects? The changing foundations of political communication. *Journal of Communication, 58*(4), 707–731.

Bennett, W. L., & Segerberg, A. (2011). Digital media and the personalization of collective action. *Information, Communication & Society, 14*(6), 770–799.

Bennett, W. L., & Segerberg, A. (2013). *The logic of connective action: Digital media and the personalization of contentious politics.* New York, NY: Cambridge University Press.

Bennett, W. L., Segerberg, A., & Knüpfer, C. B. (2018). The democratic interface: Technology, political organization, and diverging patterns of electoral representation. *Information, Communication & Society, 21*(11), 1655–1680.

Berners-Lee, Tim. (2000). *Tejiendo la red.* Madrid, Spain: Siglo Veintiuno Editores.

Bernete, F. (2013). Identidades y mediadores de la ciudadanía digital. En F. Sierra (Ed.), *Ciudadanía, Tecnología y Cultura: Nodos conceptuales para pensar la nueva mediación digital* (pp. 151–179). Barcelona, Spain: Gedisa.

Boulianne, S. (2015). Social media use and participation: A meta-analysis of current research. *Information, Communication & Society, 18*(5), 524–538. https://doi.org/10.1080/1369118x.2015.1008542.

Castells, M. (1996). *The information age: Economy, society and culture, Volume I: The rise of the network society.* Cambridge, UK: Blackwell.

Contreras, F. R. (2013). La colaboración en la esfera pública digital. En F. Sierra (Ed.), *Ciudadanía, tecnología y cultura: nodos conceptuales para pensar la nueva mediación digital* (pp. 119–149). Barcelona, Spain: Gedisa.

Cotarelo, R. (2010). *La Política en la Era de Internet.* Valencia: Tirant lo Blanch.

Dader, J. L. (2009). Ciberpolítica en los Websites de Partidos Políticos. La experiencia de las Elecciones de 2008 en España ante las Tendencias Transnacionales. *Revista de Sociología Política, 17,* 45–62.

Dahlgren, P., & Álvares, C. (2013). Political participation in an age of mediatisation. *Javnost-The Public, 20*(2), 47–65.

Dalhgren, P. (2005). The Internet, public spheres, and political communication: Dispersion and deliberation. *Political Communication, 22*(2), 147–162.

De Marco, S., Robles, J. M., & Gómez, D. (2019). *Connective parties and communicative political compromises: The lack of real interaction.* Political Science Meeting, Brussels.

Deleuze, G., & Guattari, F. (1983). *Anti-Oedipus.* Minneapolis: Minnesota University Press.

Deuze, M. (2006). Participation, remediation, bricolage: Considering principal components of a digital culture. *The Information Society, 22*(2), 63–75.

Fraticelli, D. (2013). La clave está en la desintermediación. Entrevista a Hugo Pardo Kuklinski. *Letra, Imagen, Sonido, Ciudad Mediatizada, 10,* 95–104.

Galindo, J. L. (2013). Comunicología e ingeniería en comunicación social del servicio de redes sociales Facebook. De la arquitectura a la ingeniería de la cultura y la cibercultura. En S. F. Ciudadanía (Ed.), *Tecnología y Cultura: Nodos conceptuales para pensar la nueva mediación digital* (pp. 285–311). Barcelona: Gedisa.

Gálvez Muñoz, L. A. (2009). Aproximación al Voto Electrónico Presencial: Estado de la Cuestión y Recomendaciones para su Implementación. *Teoría y Realidad Constitucional, 23,* 257–270.

Gibson, W. (1984). *Neuromancer.* New York: Ace Books.

Gómez, I. S. (2004). Hacia un arte desterritorializado: La influencia de Internet en el proceso de globalización del arte. *Art ifilosofia, 38,* 183–190.

Hague, B., & Loader, B. D. (1999). *Digital democracy: Discource and decision-making in the information age.* London, UK: Routledge.

Islas, O., & Arribas, A. (2010). Comprender las redes sociales como ambientes mediáticos. In A. Piscitelli (Ed.), *Facebook y la postuniversidad: sistemas operativos sociales y entornos abiertos de aprendizaje* (pp. 147–163). Barcelona, Spain: Ariel.

Jenkins, H. (2006). *Convergence culture: Where old and new media collide.* New York: New York University Press.

Jeong, C. H. I. (2007). *Fundamental of development administration.* Selangor, Malaysia: Scholar Press.

Karakaya, R. (2005). The Internet and political participation: Exploring the explanatory links. *European Journal of Communication, 20,* 435–559.

Keck, M. E., & Sikkink, K. (1999). Transnational advocacy networks in international and regional politics. *International Social Science Journal, 51*(159), 89–101.

Koh, C. E., & Prybutok, V. R. (2003). The three-ring model and development of an instrument for measuring dimensions of e-government functions. *Journal of Computer Information Systems, 33*(3), 34–39.

Lago, S., & Marotias, A. (2007). Los movimientos sociales en la era de Internet. *Razón y Palabra, 54.* http://www.razonypalabra.org.mx/anteriores/n54/lagomarotias.html.

Lemos, A. (2002). *Cibercultura, Tecnologia e Vida Social na Cultura Contemporánea*. Porto Alegre, Brazil: Sulina.

Loader, B. D., & Mercea, D. (2011). Networking democracy? *Information, Communication & Society, 14*(6), 757–769.

Neuman, W. R., Bimber, B., & Hindman, M. (2011). *The Internet and four dimensions of citizenship: The Oxford handbook of American public opinion and the media* (pp. 22–42). Oxford, UK: Oxford Handbook.

Papacharissi, Z. (2010). *A networked self: Identity, community, and culture on social network sites*. New York, NY: Routledge.

Peinado, F. (2011, December 29). "Somos el 99%", una frase para el recuerdo de 2011. *BBC.com*. Retrieved from http://www.bbc.com/mundo/noticias/2011/12/111220_eslogan_99_por_ciento_stiglitz_fp.

Rheingold, H. (2002). *Smart mobs: The next social revolution*. Cambridge, UK: Basic Books.

Robles, J. M., De Marco, S., & Molina, O. (2010). La e-administración como modelo de democracia digital débil. *Revista Española de Sociología., 14*, 15–29.

Robles, J. M., De Marco, S., & Antino, M. (2013). Activating activist: The links between political participation and digital political participation. *Information, Communication and Society, 16*(6), 856–877.

Robles, J. M., De Marco, S., & Antino, M. (2014). Digital skills as a conditioning factor for digital political participation. *Communications, 39*(1), 43–65.

Sampedro, V. (2006). ¿Redes de nudos o vacíos? Nuevas tecnologías y tejido social. *Documentación social, 140*, 25–38.

Sampedro, V., López, J. A., & Muñoz, C. (2012). Ciberdemocracia y cibercampaña: ¿Un matrimonio difícil? El caso de las Elecciones Generales en España en 2008. *Arbor: Ciencia, Pensamiento y Cultura, 188*, 657–672.

Sánchez, J. M., & Magallón, R. (2016). Estrategias de organización y acción política digital. *Revista de la Asociación Española de Investigación de la Comunicación, 3*, 9–16.

Scolari, C. (2008). *Hipermediaciones. Elementos para una Teoría de la Comunicación Digital Interactiva*. Barcelona, Spain: Gedisa.

Subirats, J. (2015). Todo se mueve. Acción colectiva, acción conectiva. Movimientos, partidos e instituciones. *Revista Española de Sociología, 24*, 123–131.

Van Laer, J., & Van Aelst, P. (2010). Internet and social movement action repertories. *Information, Communication & Society, 13*(8), 1146–1171.

Verba, S., Nie, N., & Kim, J. (1978). *Participation and political equality: A seven-nation comparison*. London, UK: Cambridge University Press.

Villanueva-Mansilla, E. (2016). Acción conectiva, acción colectiva y medios digitales: posibilidades para la comunicación política en los tiempos de Internet. *Contratexto, 24*, 57–76.

Virilio, P. (1986). *Speed and politics*. New York: MIT Press.

CHAPTER 5

The Dreams of Technological Reason Generate Monsters

The decade of the nineteen nineties was marked by a notable disinterest in politics, and many academics entered into debates on the possibility of defining a democratic system in which citizens could attain a more prominent role and which would strengthen their ties to public institutions. It was hoped that all citizens, or at least a large number of them, could take part in discussions on topics of social, political and economic significance that affected them directly (Barber, 1984; Cohen, 1996; Elster 2001; Putnam, 1995; Robles, 2011).

Since that time, and from this perspective, citizen participation has had two aspects, first of all it is *democratic*, because it pursues collective decision making and seeks to involve all those interested, and secondly it has a *deliberative dimension* because the discussion requires deliberation of the arguments (Elster, 2001). It was expected that citizens could express their

---

"The dreams of reason" is an engraving by the Spanish painter Francisco de Goya. In it, a man is represented surrounded by books who, while asleep, dreams with animals that observe and approach him with the intention of attacking him. This work explores the idea that when enlightened reason has a dark and threatening side. Let's say that the dream of the reason is permeated with threatening monsters.

© The Author(s) 2019
J. M. Robles-Morales and A. M. Córdoba-Hernández,
Digital Political Participation, Social Networks and Big Data,
https://doi.org/10.1007/978-3-030-27757-4_5

preferences about the issues that affect them through reasoning and dialogue (Robles, 2011).

The political crisis that many countries went through during the first decades of the twenty-first century was no more than a rejection of the citizen, not only to the practices of their respective governments, but also to how the liberal democratic system itself was organized. A system hijacked by large international corporations and in which national interests were, most of the time, subject to the needs and imperatives of the global economy.

The first studies about politics and the Internet appeared in parallel with this discussion and they said that this medium would become an arena in which citizens could express their political opinions and drastically reduce this public disinterest in democratic systems. The idea of digital democracy expressed this situation to a large degree. Authors such as Hague and Loader (1999) or Van Dijk and Hacker (2001) pointed to the Internet as a suitable vehicle for the revival and reinforcement of the liberal democratic approach to politics in the twenty-first century.

The later emergence of tools for interactivity with Web 2.0 added more strength to the cause of cyberoptimism. Concepts such as disintermediation reinforce the power of citizens to increase their opportunities for political communication, as well as directly addressing and interacting with their representatives. Digital technologies also allow citizens to create cultural and political content and acquire meaningful prominence in public debate.

However, nearly two decades after the publication of these inaugural works, in the midst of the debate about disintermediation, some experts are becoming alarmed about certain negative effects that the Internet is having on political communication (Gentzkow, 2016). These effects, including polarization, incivility and homophily, are established and now attract the attention of experts due to the risk these attitudes pose for the *deliberative dimension of democracy* and, by extension, the correct role that public opinion ought to play in today's democracy (Sunstein, 2017).

If the technologies have been the solution, a tool that could revive the civic commitment of citizens, as well as structure a more inclusive and critical public opinion, the monsters of this reason do nothing but distance that possibility. Thus, together with the shadows that have been pointed out in the previous chapter, we are interested in maintaining the risks related to the way in which they relate the different agents of the digital public space. Some forms that, in many cases, are marked by confrontation, insult and exclusion. In this section we will present how these monsters of technological reason that threaten digital public opinion are structured and defined.

5.1   Homogeneity and Homophily

We are all aware that in one way or another, our contacts and networks of interactions tend to consist of people we have things in common with. The things that attract us to them are the basis for friendship, while those that repel us are usually the reason for distance and even enmity.

This tendency that we experience in our daily lives has been termed *Homophily* in scientific literature (Kadushin, 2012). According to this idea, people with common attributes such as the same taste or aspirations have a higher probability of establishing bonds of friendship or association.

Previously, when Lazarsfeld and Merton (1978) examined this topic, they asserted that there were two types of homophily: *status homophily* and *value homophily*, also known as homogeneity, each with specific characteristics. The first is based on personal attributes such as age, race or sex, or acquired as a product of our personal achievements. Other factors that may play a part could include our employment status, our education or our marital status. This is something that we have all learned from our own experiences. For example, our networks of friendships consist of people who are more or less of the same age and/or whole cultural or educational level is similar to ours (*status homophily*). Naturally, this behaviour is not universal, because we all have friends who are clearly different from us as well, but what we have described above can be regarded as general behaviour.

The second type of homophily is called *homogeneity* and is based on more subjective qualities. In these cases, it is not our personal, social or class attributes that matter, but the fact that we share some common values. According to this meaning, emotional attitudes and dispositions can be generated or adapted as a result of the relationships that form between people (Erickson, 1988). There are therefore two underlying models for homogeneity. One the one hand, there are shared standards or values that can bring people together to share information, friendship, support, etc. On the other there are shared attributes such as age or education that can lead us to establish common values, standards and emotional postures (Kadushin, 2012).

These general characteristics for *homophily* and *homogeneity* also apply to digital social media such as *Twitter* and its political dimension. Vaccari et al. (2016) indicate that we can distinguish between three types of communicative structures when discussing political homophily on social media.

First, there are the "support networks" in which people interact with others who share their opinions. Secondly, there are "opposition networks" in which individuals can contrast their point of view with the views of people who think differently, and thirdly, there are "mixed networks" in which users interact with people having similar and contrasting outlooks at the same time (Nir, 2011). Homophily will therefore only occur in the first scenario while the opposition and mixed networks represent other forms of communication.

In recent years, a number of studies have shown that *homophily* is one of the main consequences of the political development of digital social media (Bakshy, Messing, & Adamic, 2015). The most widely accepted interpretation of this is that the Internet has increased people's ability to choose their sources, making ideological self-segregation much easier (Sunstein, 2017).

This thesis has been challenged by other interpretations that, although less widely accepted, have been the focus of interest by experts. For example, Holbert, Garret, and Gleason (2010) affirm that while Internet users are able to select people and content that increase the possibility of being exposed to messages of reinforcement, they do not necessarily lead to the avoidance of discordant opinions. Similar arguments are raised by Wojcieszak and Mutz (2009) when they state that accidental exposure to political content online can act as a countermeasure that increases contact with opposing political views.

Despite this, a number of empirical studies have helped us to outline the way in which homophily affects political communications by revealing the differences between some citizens and others depending on their individual characteristics. For example, each person's off-line personal network and the number of political messages exchanged will correlate positively with the probability of involvement in varied political networks on digital platforms. The most active Twitter users with the most followers prefer to interact with users having a similar political status (Himelboim et al., 2016). In connection with this behaviour, bloggers who are interested in political issues tend to publish links with a similar orientation to their own (Lawrence, Sides, & Farrell, 2010).

At this point it is important to add some considerations about homophily and its political effects. Homophily is, to an extent, an inevitable result of social interaction. Empirical results tell us that the great majority of citizens tend to associate with others who share their values and ideology. We should note that this behaviour is, in rational terms, easy to understand. So, for example, the information received from people who share our opinions can be seen as more reliable and truthful than that which comes from people who think differently. Similarly, limiting ourselves to information from sources that back up our position is a more efficient way of acquiring information because we need not spend as much time and effort in verifying it (Downs, 1957).

This rational interpretation of homophily has led some writers to show how this type of behaviour encourages the generation of collective action processes by coordinating critical voices that share the same orientation (Centola, 2013). When it comes to electoral campaigns, we would expect people with similar attitudes and characteristics to relate to each other. The risk, naturally enough, is that the level of homogeneity can reach such heights that the political views shared can become so shuttered that they prevent any kind of approximation to alternatives, and the group comes to define itself by its opposition to others. In this case, what we are seeing is polarization and, as we shall see in the next section, that is a different scenario.

5.2   Polarization

The debate about whether the Internet, and more specifically digital social media, generates political polarization is an acutely contemporary one. Authors like Sunstein (2017) claim that the political use of the Internet may polarize political debate on social networks by reinforcing the convictions citizens already hold, and by limiting their interaction with people holding different political positions. Other academics, in contrast, such as Barbera (2014), aim to show that exposure to politically diverse online content has a positive effect on moderating the political attitudes of Internet users.

We can also find another perspective in academic literature to complement these two views arguing that polarization depends largely on how the subjects studied are grouped or classified. Some studies have found, for example, that polarization is more common among more informed citizens (Boxell, Gentzkow, & Shapiro, 2017), among the young (Lelkes, 2016) and among those who are most interested in politics (Davis & Dunaway, 2016).

The fact is that there is no consensus about the existence of political polarization on social networks, nor is there a single definition of the idea. Some, like Abramowitz (2010) see it as a process that makes the attitudes and opinions of citizens and parties stronger and more consistent; others, such as Fiorina and Samuel (2008) stress the distance between people's points of view in a specific political context. Nor is there unanimous agreement about which factors to observe to say whether polarization exists in a digital environment or not. Studies focus on issues such as the actions of political parties in the context of polarization (Darmofal & Strickler, 2019) the more or less impolite language used by those taking part (O'Sullivan & Flanagin, 2003) or the emotional and affective reactions of citizens to certain political events (Lelkes, 2016).

There is, however, a certain degree of consensus about whether polarization represents a threat for political communication to the extent that it positions citizens on potentially conflictive issues and prevents discussion that could lead to consensus or understanding, creating factions that could be seen as negative for political processes (Abramowitz, 2010).

The growth of political participation on digital networks and the passion with which certain topics are discussed have led some authors to hold that the network has done nothing more than underline the differences between citizens. Mason (2015), for example, has shown how the political use of the Internet creates politically biased "gangs" that further break up political debate. Social networks are now being incorporated into political strategies and form part of the repertoire of debate mechanisms, but this creates a risk of encouraging even more polarization in the political arena.

We know that the levels of citizen polarization are closely linked to membership in certain social groups and the consumption of some kinds of political information (Boxell et al., 2017). This circumstance could be a consequence of the polarization of the main political agents: parties, organizations, the media, etc. We are therefore looking at a phenomenon of contagion in which the citizens who take part in debates on social media adopt different positions when reflecting on the positions of their main political reference points, either because they reply to them or because they share the same content.

The danger of this contagion effect is what Lelkes (2016) calls "affective polarization", a process in which citizens tend to radicalize their emotions and/or feelings about different topics, following the polarized speeches of political parties and public representatives. This radicalization of citizens would only serve to increase the hostility between the parties (Haidt & Hetherington, 2012).

There are also studies showing how citizens tend to adopt more extreme attitudes and positions in online political debates, not only in relation to the position of the most important actors, but also in line with the positions of other citizens, or to the extent to which the debate touches more on people, such as politicians, rather than on the central issues of public concern (Herbst, 2010).

In this book, we are particularly interested in this subjective dimension of polarization. Specifically, we would like to find evidence of the recurrence or not of "affectively polarized" positions (Lelkes, 2016) in a context of political communication focused on the debate on Twitter about the Unidos Podemos candidacy for the 2016 Spanish presidential elections. In this case we would like to use two variables that are rarely measured, and which in our opinion, can significantly improve our knowledge of polarization in general, and "affective polarization" in particular. These variables are the degree of involvement in online political debate by political agents and the debate topics.

We therefore aspire to discover the following circumstances. First of all, to know whether, in the case of the study we are looking at, we are actually in a context marked by "affective polarization". In our opinion, this context would be one where the agents that take part in a political debate show polarized emotions, in our case, positive or negative ones. We are also interested in knowing whether, in the presence of "affective polarization", this is more or less present among agents who are fully, barely or moderately engaged in the debate. Finally, we would like to know what relation exists between the topic of debate and the affective positions of the political agents. The results of this research are essential to measure the democratic quality of the online debate on the candidacy of Unidos Podemos in the Spanish general elections and, therefore, to improve our understanding of the use of social media in this type of political process.

5.3   Incivility

The existence of political incivility, understood as the use of disrespectful or offensive language in political debate or discussions, is hardly new. There are countless examples of this phenomenon throughout history. Even so, there are few figures who can claim as many notable insults as the former British prime minister Winston Churchill who, in one of his most memorable expressions referred to the labour leader Stafford Cripps in the following terms: "He has all the virtues I dislike and none of the vices I admire". Churchill is also credited with a less admirable retort to Bessie Braddock. The Labour Member of Parliament accused Churchill of being drunk when attending a public event, to which he is claimed to have responded: "My dear, I may be drunk and you are ugly, but tomorrow I shall be sober".

Any list of today's exchanges would be a long one. One debate which was broadcast nationwide in Spain during the election campaign of 2016 saw the Socialist Party candidate address the President Mariano Rajoy, saying: "You are not a decent person". The president then responded on live television, before millions of viewers perplexed by what they were seeing: "you are vile, mean and despicable."

Many believe that digital political communications have made this terrible legacy of analogue debate even worse. One famous example would be the tweet by the then candidate for the Presidency of the United States, Donald Trump, who said of his political opponent Hillary Clinton: "If Hillary Clinton cannot satisfy her husband, what makes her think she can satisfy America". The former president of Italy, Silvio Berlusconi, commented on the French president and his wife Brigitte Trogneaux by saying: "Macron is a bright kid who has been lucky enough to find a good mother to look after him".

Can we really say that digital communications have made this type of message more extreme? What exactly is political incivility? What do all these behaviours have in common? The experts have not come to an agreement on this. There are, on one hand, a series of studies that define incivility as when the actions of one of the participants in a debate are expressed in a rude manner. We should then consider that someone is uncivil in their behaviour if they use vulgar, insulting or ironic language (Herbst, 2010). It would appear that talking ironically about the vices and virtues of a political opponent, as in the first quote from Winston Churchill above, could be classed as incivility under this definition.

The use of highly emotional speech or expressions, whether provoked by a political adversary, a stressful situation or simply the communicative incapacity of the speaker, is also considered a lack of civility (Sobieraj & Berry, 2011). We could therefore say that a person is uncivil when they "lose their temper" and get emotionally carried away from rational and balanced behaviour. In the case we mentioned above, the president of the Spanish government was unable to control himself on receiving an openly hostile comment from his political opponent.

Naturally, insults and contempt are also seen by the experts as a display of uncivil behaviour (Coe, Kenski, & Rains, 2014). Calling someone ugly would surely be seen as acting in poor taste. However, to do so in public, in an official, political context adds factors that go beyond poor taste and make the comment, at the very least, rude.

Even so, some authors refuse to consider this type of comment or expression as incivility. For example, Papacharissi (2004) agrees that insults, sarcasm and irony are clearly signs of impoliteness or a lack of courtesy, but not of *incivility*. For her, this type of behaviour is a threat to democracy and several of its most fundamental principles, such as inclusive, equal and open debate.

We could consider that impolite language can be seen, in etymological and philosophical terms, as incivility only if we can affirm that a person or comment is not civilized. By highlighting this aspect, we are specifically saying that the comment is not appropriate to the rules of courtesy or civility. Papacharissi (2004), however, is correct to say that, in political terms, *incivility*, when understood as rudeness, is less decisive than *incivility* interpreted as a threat to democracy.

When President Trump says that his opponent is not qualified to govern the country because she cannot satisfy her husband, not only is he insulting Hillary Clinton (incivility as a lack of courtesy), but also breaking the rules of political debate and consideration. The comment also offers a stereotyped view of women, defined as a social group whose job is to take care of their husbands, removing them from public life. Similar ideas arise when we consider Berlusconi's message. It is an insulting comment that also questions the President of France's maturity and independence. Once again, the comment dismisses the opponent from public debate by claiming that his ability is derived from his wife and is not his own, as a free and independent person.

We agree with Papacharissi that incivility becomes a problem for democracy when the messages expressed seek to disqualify people or groups from the public arena and to undermine the rules of debate. At this point, however, we need to ask to what degree the Internet and social media have aggravated these types of attitudes and behaviour. The classic, most widespread argument is that the Internet is effectively becoming a space in which people show less respect in the expressions they use and in their relations with other people.

This is by no means a recent phenomenon. The term *Netiquette* was coined at the end of the last century to define the behaviour that users were expected to conform to in conversation. It was introduced because the increase in computer-based communications had already been marked by offensive language and attitudes in this early stage. It was an attempt to ensure that the forms of courtesy that prevailed in off-line communications would transfer to the online realm.

The massive growth of social networks was accompanied by a series of studies that offered key points for interpreting this phenomenon. Experts showed that the anonymity that was a feature of digital communications was often a stimulus for antisocial behaviour (Rains, Kenski, Coe, & Harwood, 2017). They also observed that as we increasingly interacted with strangers, people with whom we had only weak bonds, this tended to make people less inhibited, leading to more uncivil behaviour among Internet users (Gervais, 2015). The "anonymity" factor increasingly appears to be a central issue in understanding the potential increase in online incivility. In fact, as Rowe (2015) pointed out, if we move online debates from spaces that are fully anonymous (newspaper comment sections) to others such as Facebook in which identities are public, the number of impolite comments and expressions goes down.

Even so, the content and profile of the user is also relevant. As Gervais (2015) said, incivility is a self-generating process. Exposure to highly uncivil messages increases the likelihood that this type of message will then be used by other people. It would appear that granting acceptance of bad attitudes encourages people to indulge in their less positive traits. A similar effect is also seen when the uncivil message is issued by a celebrity or source of political authority. Boxell et al. (2017) showed that the debates arising from the uncivil tweets of a public figure lead to the rudest and most offensive messages.

We have plenty of reasons to believe that, on the one hand, political communication is not free of unacceptable behaviour and commentary, and that this attitude reflects a problem of courtesy and politeness, but most of all it represents a significant deterioration of the proper conditions in which political communication should take place in democracy. Reasoning from this, experts indicate that the nature of digital technology is amplifying a pre-existent tendency.

5.4   Flaming

Flaming is a type of behaviour that is often found in online political debate that is characterized by moving beyond uncivil expressions and behaviour in order to disrupt the debate and make it impossible to continue. The literature has also used the expression *troll* to define people who, for whatever reason, try to provoke anger in public debate by spreading discord and hatred. So, *flaming* refers to the process while *troll* would refer to the person who engages in this activity.

The disruption of online conversations, however, does not always arise because of the intentional action of one of the persons involved. Institutions, groups and governments increasingly resort to robots, commonly referred to as *bots* to try to influence public opinion in political processes as important and election campaigns and other debates on social issues that are politically significant.

Unfortunately, there are many different examples of this interference in the free exercise of public opinion. A lot has been said about the actions of Russian agents in the 2016 United States presidential election campaign, or in the debate about Catalan independence from Spain in 2017. In these cases, all the parties involved in the political process had serious doubts about whether the information that was doing the rounds on the social networks was true or not, or if it was intended to generate anger and to inflame political debate.

In this sense, the so-called *fake news*, which is pseudo-journalism spread on social networks with the aim of deliberately misinforming or misleading, becomes the perfect complement to the action of the *bots* and the *trolls*. In other words, it is misinformation or "doctored" news as a way to inflame public debate.

Naturally, this situation has opened a debate on the manipulation of public opinion in digital spaces. The social network Twitter recently decided to take matters in hand in view of the risk of these fraudulent processes affecting its credibility as a vehicle for political communications. In 2018, the social network destroyed some of the trolls of the Government of Mauricio Macri. It would appear that Twitter tired of the manipulation exercised through mass messages sent by robots working for Argentina's highest institution and banned them.

Besides the effect that these practices could have on the business goals of social networks such as Twitter, the fact is that they drag the political sphere down to a point at which trust in the information and the intentions of the publisher is seriously compromised. In this sense, recent events have proved that Morozov (2011) was right to warn about the interference of established powers in the configuration of digital public space. According to this author, Internet is far from being an autonomous, critical space, but instead a place where information is distorted by organizations that co-opt opinion either openly, or by using legitimate political techniques, and make it politically heteronomous. Naturally, this view represents a serious challenge to the aspirations to transform this technology into an alternative space free from interference.

References

Abramowitz, A. I. (2010). *The disappearing center: Engaged citizens, polarization, and American democracy*. New Haven, CT: Yale University Press.

Bakshy, E., Messing, S., & Adamic, L. A. (2015). Exposure to ideologically diverse news and opinion on Facebook. *Science, 348*(6239), 1130–1132.

Barber, B. (1984). *Strong democracy: Participatory politics for a new age*. Berkeley: University of California Press.

Barbera, P. (2014). *How social media reduces mass political polarization: Evidence from Germany, Spain, and the US* (p. 46; Job Market Paper). New York University.

Boxell, L., Gentzkow, M., & Shapiro, J. M. (2017). *Is the Internet causing political polarization? Evidence from demographics*. Cambridge: National Bureau of Economic Research.

Centola, D. (2013). Social media and the science of health behavior. *Circulation, 127*(21), 2135–2144.

Coe, K., Kenski, K., & Rains, S. A. (2014). Online and uncivil? Patterns and determinants of incivility in newspaper website comments. *Journal of Communication, 64*(4), 658–679.

Cohen, J. (1996). Deliberation and democratic legitimacy. In A. P. Hamlin & P. Pettit (Eds.), *The good polity*. Oxford, UK: Blackwell.

Darmofal, D., & Strickler, R. (2019). Introduction. In *Demography, politics, and partisan polarization in the United States, 1828–2016* (Vol. 2). Spatial demography book series. Cham: Springer.

Davis, N. T., & Dunaway, J. (2016). Party polarization, media choice, and mass partisan-ideological sorting. *Public Opinion Quarterly, 80*(S1), 272–297.

Downs, A. (1957). *An economic theory of democracy*. New York, NY: HarperCollins.

Elster, J. (comp.). (2001). *La democracia deliberativa*. Barcelona, Spain: Gedisa.

Erickson, B. H. (1988). The relational basis of attitudes. *Social Structures: A Network Approach, 99*(121), 443–475.

Fiorina, M. P., & Samuel, J. A. (2008). Political polarization in the American public. *Annual Review of Political Science, 11*, 563–588.

Gentzkow, M. (2016). *Polarization in 2016* (Toulouse Network of Information Technology White Paper).

Gervais, B. T. (2015). Incivility online: Affective and behavioral reactions to uncivil political posts in a web-based experiment. *Journal of Information Technology & Politics, 12*(2), 167–185.

Hague, B., & Loader, B. D. (1999). *Digital democracy: Discourse and decision-making in the information age*. London, UK: Routledge.

Haidt, J., & Hetherington, M. J. (2012). Look how far we've come apart. *The New York Times*.

Herbst, S. (2010). *Rude democracy: Civility and incivility in American politics*. Philadelphia, PA: Temple University Press.

Himelboim, I., Sweetser, K. D., Tinkham, S. F., Cameron, K., Danelo, M., & West, K. (2016). Valence-based homophily on Twitter: Network analysis of emotions and political talk in the 2012 presidential election. *New Media & Society, 18*(7), 1382–1400.

Holbert, R. L., Garrett, R. K., & Gleason, L. S. (2010). A new era of minimal effects? A response to Bennett and Iyengar. *Journal of Communication, 60*(1), 15–34.

Kadushin, C. (2012). *Understanding social networks: Theories, concepts, and findings.* New York, NY: Oxford University Press.

Lawrence, E., Sides, J., & Farrell, H. (2010). Self-segregation or deliberation? Blog readership, participation, and polarization in American politics. *Perspectives on Politics, 8*(1), 141–157.

Lazarsfeld, P. F., & Merton, R. K. (1978 [1955]). Friendship as a social process: A substantive and methodological analysis. In M. Berger, T. Abel, & C. H. Page (Eds.), *Freedom and control in modern society* (pp. 18–66). New York, NY: Octagon Books.

Lelkes, Y. (2016). Mass polarization: Manifestations and measurements. *Public Opinion Quarterly, 80*(S1), 392–410.

Mason, L. (2015). "I disrespectfully agree": The differential effects of partisan sorting on social and issue polarization. *American Journal of Political Science, 59*(1), 128–145.

Morozov, E. (2011). *The net delusion: How not to liberate the world.* London: Penguin Books UK.

Nir, L. (2011). Disagreement and opposition in social networks: Does disagreement discourage turnout? *Political Studies, 59*(3), 674–692.

O'sullivan, P. B., & Flanagin, A. J. (2003). Reconceptualizing 'flaming' and other problematic messages. *New Media & Society, 5*(1), 69–94.

Papacharissi, Z. (2004). Democracy online: Civility, politeness, and the democratic potential of online political discussion groups. *New Media & Society, 6*(2), 259–283.

Putnam, R. (1995). Bowling alone: America's declining social capital. *Journal of Democracy, 6,* 65–78.

Rains, S. A., Kenski, K., Coe, K., & Harwood, J. (2017). Incivility and political identity on the Internet: Intergroup factors as predictors of incivility in discussions of news online. *Journal of Computer-Mediated Communication, 22*(4), 163–178.

Robles, J. M. (2011). *Ciudadanía digital: Una introducción a un nuevo concepto de ciudadano.* Barcelona: Editorial UOC.

Rowe, I. (2015). Civility 2.0: A comparative analysis of incivility in online political discussion. *Information, Communication & Society, 18*(2), 121–138.

Sobieraj, S., & Berry, J. M. (2011). From incivility to outrage: Political discourse in blogs, talk radio, and cable news. *Political Communication, 28*(1), 19–41.

Sunstein, C. (2017). *#Republic: Divided democracy in the age of social media*. Princeton, NJ: Princeton University Press.

Vaccari, C., Valeriani, A., Barberá, P., Jost, J. T., Nagler, J., & Tucker, J. A. (2016). Of echo chambers and contrarian clubs: Exposure to political disagreement among German and Italian users of Twitter. *Social Media+Society, 2*(3), 1–24.

Van Dijk, J., & Hacker, K. L. (2001). *Digital democracy: Issues of theory and practice*. Beverly Hills, CA: Sage.

Wojcieszak, M. E., & Mutz, D. C. (2009). Online groups and political discourse: Do online discussion spaces facilitate exposure to political disagreement? *Journal of Communication, 59*(1), 40–56.

PART II

Disintermediation in Social Networks

In the first part of the book, we have defined the central concepts and processes to the argument of this book. These concepts are taken from the specialized literature, as well as from works of authors widely analysed and accepted. These key concepts are disintermediation, as well as their extensions in the practical life of citizens, the disintermediation of agents, spaces and messages. We had also defined the meaning of polarization, incivility, homophily and flaming. These last four concepts are, as we shall see, "the dark side of force".

This second part will have, on the contrary, a more practical and empirical dimension. Thus, we focus on several case studies to test the concepts of disintermediation of agents, spaces and messages, as well as concepts with potentially negative effects on political communication (basically polarization and incivility).[1]

For this part we have selected a set of three cases that help us to exemplify the elements, both positive and negative, that generate the disintermediation of agents, messages and spaces. The first of these is a political party, Podemos, that emerged in Spain during the political and economic crisis that began in the first decade of this millennium. The second one is a process of connective action worldwide known as Black Lives Matter that appears as a reaction to a set of abuses of authority over

---

[1] We do not transfer the general concept of disintermediation to the field of empirical research because we consider that it is a structural hypothesis and we are interested in empirically analyzing the behavior and the concrete actions of agents and persons.

African-American people in the United States. Finally, the third case is Bring Back our Girls. It is a campaign to make visible and rescue a group of girls kidnapped by a terrorist group in Nigeria.

The three concepts that will be exemplified through these cases are "ideal types". That is, conceptual instruments that allow us to synthesize reality to interpret and study it. In practice, on the ground, these types naturally do not occur in a "pure form". That is, there is no case study that would allow us to isolate the idea, for example, of disintermediation of the agents. On the contrary, any of the cases presented here have a bit of each of the three types of disintermediation and, if we choose them to exemplify a specific type, it is only because in terms of exposition it has seemed more intuitive and clearer to us. Reality does not fit the sociological categories, although at times, sociologists strive, stubbornly, to do so. I hope the reader shares this decision.

We have taken the decision to distribute the cases of study among representative processes of what we have called conventional and unconventional political participation. Thus, in this part, we have a case study, the Political Party Podemos, which represents conventional political participation. Likewise, we have two cases of unconventional political participation: Black Lives Matter and Bring Back our Girls. There is no intention to generate content that allows us to speculate about the differences between one type of participation and another. This is not the objective and the empirical material is not sufficient for this purpose. The idea of including cases as diverse as possible to represent the complexity and extension of the processes that are being shown here.

Unlike the next part, in this, we have used commercial tools to analyse the data of our case studies. With this, we want to show that the analysis of social networks and the communication processes that take place there are not only available to experts in data analysis, but also the researchers less trained in the use of mathematical tools and Advanced computer systems can also be introduced in this type of analysis.

Naturally, this type of tool has the "black box" effect. That is, they do not allow knowing how information is being processed, as well as the methodology and algorithms that the tool uses. Likewise, these tools are not adapted to the objectives and needs of the researcher, but rather, they must be content with what the tool offers. It is for this reason that we recommend, whenever possible, to participate in multidisciplinary research groups in which computer scientists, mathematicians and social scientists can design and process information with full control of the process.

The cases selected to exemplify each of the concepts respond to a triple interest. In the first place, we had been moved by academic but also by social interest. So, cases have been selected with an important impact on global public opinion and with a "viral" character. Second, these case studies arise in diverse contexts and countries affected by diverse circumstances. Thus, we emphasize populism and negative partisanship in the 2006 electoral campaign in the United States or the emergence of a connective party like Podemos in the elections of the same year in Spain. We focus on cases framed in electoral processes, but also in contexts of open and spontaneous public debate. Finally, this book deals with public opinion and political communication. Therefore, the case studies are examples of political communication and public debate.

CHAPTER 6

The Disintermediation of the Agents: The Case of #UnidosPodemos

6.1   In Memoriam Los Indignados

On 17 December 2010, Mohamed Bouazizi, a 26-year-old salesman from Sidi Bouzid, in the centre of Tunisia, set fire to himself in front of the government building as a desperate protest against a system under which the state authorities subjected him and his compatriots to continuous abuses (Castells, 2012).

A few hours later, the country entered a phase of collective action that lasted for months. Bouzazi's protest lit the spark of revolts that spread across 15 of the 22 countries that make up the Arab world. In all of these, the absence of democracy, harsh economic conditions and the lack of opportunities for the future, coupled with the states' violations of human rights all contributed to the uprising that we would later come to call the *Arab Spring*.

By the end of 2011, the knock-on effect was undeniable, and dominoes continued to fall across the north of Africa: the Yemeni leader Alí Abdulá Saleh resigned in Riyadh on 23 November and became the fourth head of state to be removed by civil mobilizing, following Ben Alí (Tunisia), Mubarak (Egypt) and Gaddafi (Libya).

Meanwhile, another branch of social protest, that of the Occupy movement spread around the world. On 15 May 2011, the citizens of Madrid (Spain) occupied one of the most symbolic central squares in the city, the Puerta del Sol, in protest to the policies of the Spanish Government in

© The Author(s) 2019
J. M. Robles-Morales and A. M. Córdoba-Hernández,
Digital Political Participation, Social Networks and Big Data,
https://doi.org/10.1007/978-3-030-27757-4_6

the middle of a deep economic crisis. While the causes, context and motivations were very different from those of the Arab countries, the media quickly came to link them together and, over time, reveal them to be distinct aspects of the same phenomenon, that of a worldwide uprising.

Once again, the hundreds of thousands of citizens who gradually occupied the squares of the main cities in Spain demanded more democracy and participation in public decision making. At the same time, this participatory quake began to be echoed in other parts of the world. Portugal, France, Germany, the United States, Mexico and Chile, to give only a few examples.

Nearly a decade after the start of this cycle of protest, the situation in the Arab countries is far from ideal. In Tunisia, despite having several free elections, a new constitution and advances in social issues, citizens feel that they are still a long way from the aspirations they held at that time. In Egypt, the political situation is even more unstable and human rights abuses are prompting a reduction in the financial aid that Western countries provide, in addition to other consequences. The situation in Syria is even more serious. The 2011 protests led to a civil war and purges that are oppressing the populace today.

Spain, on the other hand came out of the cycle of protest differently. Few authors challenge the consequences and achievements of this movement. The 15-M movement was founded on citizen indignation that was in tune with the framework of collective action, supporting its public assemblies and arousing popular support on a symbolic level (Antón, 2012; Díez García, 2014; Laraña & Díez, 2012; Sampedro & Lobera, 2014).

This movement invigorated Spanish civil society which has become more active and open since the middle of the nineteen nineties (Laraña, 2007, 2009; Laraña and Díez, 2010, 2012, 2013). Some authors have suggested that it is a continuation of previous militancy and activism (Fominaya, 2014; Maeckelbergh, 2012). Its broad-based character, ideological pluralism and the variety of traditions and practices it brought together enabled the 15M to extend the reach of its collective actions (Díez García, 2014).

The 15M movement revealed the possibilities for citizen participation in less visible areas that are far removed from formal politics through digital technologies. This was the expansion of new *political forms, subpolitics*, according to Beck (1992), that implied the opening of new ways to participate. The use of Internet and widely used applications such as Facebook and Twitter made it easier for messages and citizens' demands to spread and resonate. This is why authors such as Castells (2012) stressed the decisive role of these technologies in the rapid, viral distribution of new ideas and messages and favoured the contagion of these social movements nationally and internationally.

The genesis of the 15M movement was associated with the Free Culture Movement and the use of digital social networks (Candón, 2013; Castells, 2012; Toret, 2012). In early 2011, the Coordination platform *Grupos por la Movilización Ciudadana* created a Facebook group which was quickly and spontaneously joined by bloggers and people in the *Estado del Malestar* Citizen Movement and the *No les votes* movement (Elola, 2011). In the next few weeks, this group became a forum for debate and action called *¡Democracia Real Ya!* This would be one of the main slogans used in the tent city raised in Puerta de Sol.

The context was perfect, while the screens and front pages of the media were full of images of indignation from the main cities of the Arab world in the first quarter of 2011, the social networks of Spain were on fire. The *Democracia Real Ya* group was joined by organizations like *Juventud Sin Futuro* and the Platform *Afectados por la Hipoteca*, who encouraged citizens to attend a popular demonstration that had been coordinated and organized on social media and meetings and assemblies in the different districts of Madrid, all coming together in demonstration of 15 May 2011 and the subsequent occupation by the "indignados" in the symbolic Puerta.

15M aimed to champion a more participative democracy, as reflected in its motto: "Real Democracy Now: We are not goods in the hands of politicians and bankers". This assertive stance became widely known through the intelligent use of these technologies and coverage by traditional media. Even so, digital communication technologies were not only used to spread the information, but also to organize, coordinate and *mobilize the consensus* (Klandermans, 1984).

Time passed, and the activists did not want 15M to lose its power to motivate. In mid-2013, a debate started in the centre of the movement about whether it should be turned into a political party and take part in elections, or continue as the same type of unstructured organization it had been since its founding. Opinions were divided, because a political formation could offer them the chance of representation and capacity of manoeuvre in the context of traditional politics, but it was also clear that one of the more prominent features in the identity of the Movement was precisely its sharp criticism of the main traditional political parties. How could they be assured that transformation into a political coalition would not turn them into what they had been criticizing?

In the midst of these deliberations, the Spanish political scenario was shaken by the official emergence of a new political party, *Podemos*, led by the political scientist and university lecturer Pablo Iglesias. From its foundation, Podemos attracted a considerable number of the *indignados* who had been organizing in support groups since 2011 and calling for demonstrations against the economic situation and to protest against the traditional political elites who had been sharing power in Spain, especially the socialist *Partido Socialista Obrero Español* (PSOE) and the conservative *Partido Popular* (PP).

As Meyenberg (2017) observed, the leaders of the new political party saw 15-M as a historical opportunity to harness all forms of political dissatisfaction in Spain to embark on a radical overhaul of the system. This led them to organize as a political choice able to represent many different groups and to compete for power at the ballot boxes.

The birth of Podemos and especially the electoral success in the European elections of May 2014 revealed the weakness of the traditional assemblies of 15M, with a noticeable transfer of assembly members to the circles of the party, a process that gathered speed from September, when Podemos started its process to become a political party (Martín, 2015).

In only one year since its founding, in the general elections of 2015, Podemos emerged as the third political force in Spain, triggering a period of intense instability in the traditional two-party system of PP and PSOE. When this phenomenon was studied, some authors pointed out that the political marketing strategy of the party centred on its leader, Pablo Iglesias and his charismatic presence in the media (Peris, 2018); others noted that the populist emphasis on the leader was countered by broad support from the group (Arroyas & Pérez Díaz, 2016). In this work, however, we are not going to look at these aspects, but the form of connective organization based on social networks and digital technologies that the party has used, as we shall now see.

6.2   THE CONCEPT OF THE CONNECTIVE PARTY

In a scenario of public dislike of and citizen disinterest in traditional political organizations, such as political parties and unions, we can find a series of changes taking place in these institutions that are intended to adapt them to the structure of the new digital environment and the demand for direct participation and disintermediation. Left-wing parties in many countries have seen the emergence of a form of organization unlike the usual rigid hierarchical structures. These new parties seek to channel citizen's demands for more profound democracy and innovations in the democratic system. These new parties look beyond the search for votes and a logic of democratic representation and prefer the logic of connecting citizens with political authorities, whether through party organization or political life in general.

According to Bennett, Segenberg, and Knüpfer (2018, p. 1656) a connective party is one that replaces:

> Operations such as agenda setting and candidate selection with a combination of assemblies, coordinated through technology platforms. In some cases, these party "operating systems" have been driven by necessity, such as Bepe Grillo's Five Star Movement (M5S) in Italy, which grew rapidly to become a national political force far beyond Grillo and his followers' capacity to manage it. In other cases, parties have tried to achieve a balance between the demands for direct democracy and the need for central leadership, such as the Piratar in Iceland, Alternativet in Denmark and Podemos in Spain.

The key issue here is that the left is at least engaging in transformations that affect the parties' own structures in order to adapt to a context of championing self-expression (Inglehart & Norris, 2016) and therefore to disintermediation.

Naturally, this is an emergent trend, but it is also an example of how traditional, firmly structured organizations see the development of Internet and disintermediation as a window of political opportunity. In other words, these are channels through which to obtain a better connection between the needs and demands of the citizens, and a new way of understanding politics that is the outcome of the development of digital technologies.

These parties, who must therefore find a natural space within the national political sphere, are defining their structures in inclusive ways. They do this by becoming channels for communication with the electorate. This will naturally affect the notion of representation proposed by Pitkin (1967). Connective parties do not represent through authorization, in the sense that they do not see themselves as authorized to act on behalf of citizens. Nor do they aspire to the *rendering of accounts* or *descriptive* representation, when there is a correlation of qualities or shared characteristics between representative and represented. Their representation cannot be seen as *symbolic* (when representatives have an emotional identification with the represented) or *substantive* (when the representative is mainly concerned with the benefits and interests of a group of electors).

The important aspect of these parties is not, as Sartori (2005) suggests, reciprocity, rendering of accounts or the possibility of dismissal, but the acknowledgement of a civil society that, with the development of the digital public sphere, is in a position to make use of these organizations without strong mediation. On the contrary, they are organizations that, theoretically and potentially transmit citizens' ideas to political decision making bodies. From our point of view, this type of party offers distinct criteria for representation that is adjusted to the idea of disintermediation.

If we see representation in general as the ability to take decisions that affect the general public without the direct participation of all those affected (a form of transmission of power), *connective parties* offer something rather different. First of all, this type of party assumes the direct participation of those involved, and secondly, it cannot rely on the ability to wield transferred power. Power, as regards the objectives and strategies, is shared between the participants.

This change is also noticeable in other central institutions in the public sphere. The media generate content in a range of formats in which the presence of the audience, formerly passive, is increasingly significant. Amateur journalism, consultations of the readership and other forms of involvement aim at involving citizens in debating the political and social issues presented to them through the TV screen. However, this strategy does not achieve the same democratic reach as the processes started by some parties that we include here under the category: connective parties.

6.3   A Party that Operates in Internet

Since its founding, Podemos has sought to organize itself and get ordinary people involved through the use of digital tools. This was not something new, as we saw with the 15M movement, which aimed to organize around the use of the most socially accepted applications in Spain at that time to ensure the widest possible range of citizens. For example: the most active participants used simple text editing applications such as *TitanPad*, or applications like *Mumble* to hold assemblies and videoconferences, platforms like *Bambuser* for live streaming and alternative social networks such as $N - 1$ for its enormous collaborative potential.

A number of different projects, initiatives and tools have been tested by 15M and later by Podemos to emphasize their processes of discussion and debate, in order to take decisions collectively and democratically, such as the wide-ranging project *Democracia 4.0*,[1] and others of more limited range, such as: *Propongo, Agora Voting, Loomio, Reddit* or *AppGree.*

It was not only Podemos who wanted to introduce new forms of networked and digital organization into institutions, but other parties who shared some links with Podemos, such as: *Partido X, Barcelona en Comú, Ahora Madrid* and *Ganemos Madrid,* which received wide coverage from the mass media.

Let's return to the case of Podemos. This was a political phenomenon that after only one year, in 2015, emerged as the third major political force in Spain. According to Meyenberg (2017), the party's virtue lies in its use of different online instruments to prepare its proposals and to establish its mechanisms for voting in assembly. We could say that Podemos created a public sphere 2.0 that allowed it to maintain active and cross-sector communications with its members and supporters.

The platforms used by the party also enable citizens to engage in direct digital democracy, giving them the sensation of dynamic communications in decision making, in designing the organizational model and party strategy as well as selecting the candidates.

We can see one clear example of this in the Podemos Citizen's Assembly at Vistalegre Palace in Madrid, in October 2014. On this occasion, the party chose to use *Appgree* as its channel for discussions. This application is easy to use and was downloaded by nearly 4000 people who took part in the event. The advantage of Appgree is that it can instantly show what a large number are thinking, whatever the majority agree on. How does it work? A group of any size is convened and asked a question. Everyone answers through the application and each answer is checked by a part of the group so that all the answers are assessed. From this point, the group chooses its preferred answers, retaining those with the most votes to leave the one accepted by the majority, so that the group can reach an agreement.

First of all, the questions are simple ones about what citizens look for in the party, such as: What changes do you expect Podemos to make? When the time was up, after 3300 users took part, the most popular answer, with 86% of the votes, was: "A fairer society: employment, democratic participation and a fairer distribution of wealth".

---

[1] Designed to combine the votes of political representatives in Parliament and citizen participation through Internet voting with digital certificates.

At the same time, participants were encouraged to make their own demands to the spokespersons, so that nearly 15 thousand proposals were made in the different areas. After assigning speaking turns and analysing 260,000 votes cast via smartphones, the tool worked out which of these hundreds of opinions and proposals from users generated more activity, popularity and redundancy within the group (De Ser, 2014).

At the same time, the Party participation team, its "tech heads" continued to look for all types of functional tools for citizen empowerment to include in the organization or for the community. As De Ser (2014) claimed, social networks, participation and technology and core areas of the structure they chose and support one of the stated goals of Podemos' programme: "let the grass roots speak"

Another of the organizational tools used by Podemos is the *Reddit* social network, where the "Plaza Podemos" was set up and where "Mass Briefings" take place in which the party leaders answer questions from users as part of a bottom-up communications channel. This system operates like the *Menéame* social network, and in both the users who receive the most votes become the most popular and are highlighted on the page (Machuca, 2014).

Besides these online discussion tools, the party has also innovated with its voting system. Since its founding, the party has chosen its candidates, programmes, organizational and political strategies through votes cast via Internet using software such as *Agora Voting*, which uses a code sent to the smartphone once the user registers in the web page. People who are not connected to the Internet, however, can vote in person in places prepared by the party.

Keeping this network of communications calls for a team of voluntary workers who look for "peer to peer" communication options for citizens on networks, platforms and applications. This work has produced results. To date, Podemos is the political party with the most support on social networks, way ahead of the traditional mainstream parties as shown in Table 6.1.

These official party accounts must be supplemented by all the sector and regional accounts that are operated by party circles and assemblies, in Spain and in other parts of the world.

Finally, the communication strategy of Podemos has also been acclaimed as a great success in the media (Fernández, 2014). The goal here has been very clear since the beginning: to use the mass media to promote and reinforce the image of the party leader, Pablo Iglesias. This ex-professor

Table 6.1 Followers and fans on social media of the main political parties in Spain

Network	Podemos	PP	PSOE
Twitter (followers)	1,331,475	4.364	13.737
Facebook (fans)	1,207,917	189.97	159.37
Instagram (followers)	69.966	32.541	25.752

Source Twitter, Facebook, Instagram (30 August 2018). Own research

of Political Science combined his appearances on prime-time programmes with the recording his own television programme, *La Tuerka*, which he directed and presented.

This meant that the figure of Pablo Iglesias quickly became well-known in Spain and his political ideas were discussed and debated. As an "outsider" candidate, he needed visibility and the Social Media Team quickly saw that his appearance on TV could become a form of mass propagation by increasing the audience and boosting the impact of every piece of news about Podemos (Prieto, 2014).

More recent studies emphasize the fact that Podemos's communicative strategy, as well as its extensive use of digital social networks to link citizens to party decisions, was never completed (De Marco, Robles, & Gómez, 2019). On the contrary, the reality is that although the citizens, motivated by the feeling of closeness with the party leaders, participated and communicated with them through digital social networks, the rate of response obtained was almost nil. That is to say, this connective party did not comply with the exceptions generated on its connective nature.

In the same line, during the years 2016, 2017 and 2018, Podemos chose to launch, on a national level as well as in some of the Spanish town halls in which they governed, experiences of online public debate. However, these debates did not have a real impact on the policies and political positions that the party carried out. These circumstances, as explained in Chapter 4, have generated among the Spanish citizens a disenchantment with respect to the differences between the communication strategies of the traditional parties and the connective parties.

References

Antón, A. (2012). *Movimiento 15-M: expresión colectiva de una ciudadanía indignada*. Trabajo presentado en la Jornadas de Sociología de la Asociación Madrileña de Sociología, Universidad Complutense de Madrid, marzo.

Arroyas, E., & Pérez Díaz, P. L. (2016). La nueva narrativa identitaria del populismo: un análisis del discurso de Pablo Iglesias (Podemos) en Twitter. *Cultura, Lenguaje y representación, 15*, 51–63.

Beck, U. (1992). *Risk society: Towards a new modernity*. London, UK: Sage.

Bennett, W. L., Segerberg, A., & Knüpfer, C. B. (2018). The democratic interface: Technology, political organization, and diverging patterns of electoral representation. *Information, Communication & Society, 21*(11), 1655–1680.

Candón, J. (2013). *Toma la calle, toma las redes: El movimiento 15M en Internet*. Andalucia, Spain: Atrapasueños.

Castells, M. (2012). *Networks of outrage and hope: Social movements in the Internet age*. Cambridge, UK: Polity Press.

De Marco, S., Robles, J. M., & Gómez, D. (2019). *Connective parties and communicative political compromises: The lack of real interaction*. Political Science Meeting, Brussels.

De Ser, G. (2014). La infraestructura de Podemos vive en Internet. *El País*.

Díez García, R. (2014). Does the Spanish 15M have an ideology? Issues of method and measurement. *Athenea digital: revista de pensamiento e investigación social, 14*(3), 199–217.

Elola, J. (2011). #SPANISHREVOLUTION. El 15-M sacude el sistema. *El País*. Retrieved from http://elpais.com/diario/2011/05/22/domingo/1306036353_850215.html.

Fernández, J. (2014). ¿Quién votó a Podemos? Un análisis desde la sociología electoral. In. F. M. González (Ed.), *#Podemos. Deconstruyendo A Pablo Iglesias* (pp. 75–101). Bilbao: Deusto.

Fominaya, C. (2014). Debunking spontaneity: Spain's 15-M/Indignados as autonomous movement. *Social Movement Studies, 14*(2), 142–163.

Inglehart, R., & Norris, P. (2016). Trump, Brexit, and the rise of populism: Economic have-nots and cultural backlash (Working Paper).

Klandermans, B. (1984). Mobilization and participation: Social-psychological expansions of resource mobilization theory. *American Sociological Review, 49*, 583–600.

Laraña, E. (2007). *Asociación y modernización social. Las organizaciones voluntarias en España Lo que hacen los sociólogos, libro homenaje a Carlos Moya* (pp. 735–754). Madrid, Spain: CIS.

Laraña, E. (2009). *Is still Spain a statist society? A research perspective on organizations, reflexivity and collective action*. Berkley, CA: Institute for the Study of Social Change.

Laraña, E., & Díez, R. (2010). *Las grandes manifestaciones en la prensa y el poder de persuasión de las organizaciones cívicas. Metodología de investigación y marcos de movilización*. Madrid, Spain: Informe Técnico para el CIS.

Laraña, E., & Díez, R. (2012). Las raíces del movimiento 15-M. Orden social e indignación moral. *Revista Española del Tercer Sector, 20*, 105–144.

Laraña, E., & Díez, R. (2013). Las organizaciones reflexivas y el surgimiento de la sociedad civil en España. In *XI Congreso Español de Sociología*. Madrid, UCM, julio.

Machuca, P. (2014). Podemos: Nuevas herramientas para una nueva política. *El Huffington Post*.

Maeckelbergh, M. (2012). Horizontal democracy now: From alter globalization to occupation. *Interface, 4*(1), 207–234.

Martín, I. (2015). Podemos y otros modelos de partido-movimiento. *Revista Española de Sociología, 24,* 107–114.

Meyenberg, Y. (2017). Disputar la democracia: el caso de Podemos en España. *Revista Mexicana de Ciencias Políticas y Sociales, 230,* 221–242.

Peris, M. (2018). El tratamiento periodístico del partido político Podemos en el País y Público: un análisis desde la teoría del framing. *Revista de Comunicación, 143,* 111–134.

Pitkin, H. F. (1967). *The concept of representation*. Berkeley: University of California Press.

Prieto, P. (2014). "Podemos ¿Magia potagia?". *El Huffington Post* [en línea]. Retrieved from http://www.huffingtonpost.es/pablo-prieto-fernandez/podemos-magia-potagia_b_6121532.html.

Sampedro, V., & Lobera, J. (2014). The Spanish 15-M Movement: A consensual dissent? *Journal of Spanish Cultural Studies, 14*(4), 1–20.

Sartori, G. (2005). *Parties and party systems: A framework for analysis*. Colchester, UK: ECPR Press.

Toret, J. (2012). Una mirada tecnopolítica sobre los primeros días del #15M. En A. Alcazan, Q. Axebra, S. Levi, T. SuNotissima, & J. Toret (Eds.), *Tecnopolítica, internet y r-evoluciones*. Barcelona, Spain: Icaria.

CHAPTER 7

The Disintermediation of the Message: The Case of #BlackLivesMatter

7.1 Genesis of a Networked Movement

There have been many episodes in the history of the United States in which racial discrimination has been a constant factor. Many social movements and groups have mobilized to fight for greater human rights and equality as part of the American justice system. The most recent of these has been the Black Lives Matter movement, which sought to draw attention to and campaign for the rights of black people with a massive deployment of online and off-line actions across the country that achieved international attention through the use of social networks and the mass media. How did #BlackLivesMatter come into existence, and what made it one of the most viral hashtags of 2014 on Twitter?

Although the tag became famous after it was used as a slogan on the protest marches that followed the death of Michael Brown, in Ferguson, in August 2014, it had originated a year before, in July 2013, when George Zimmerman was cleared of the second-degree murder of Trayvon Martin, a 17-year-old African American, in Sanford, Florida.[1] At that time, Alicia

---

[1] Martin was murdered on the night of 26 February 2012. Zimmerman never denied killing him, but declared that the young man was behaving suspiciously, and that is why he shot him (Gutman & Tienabeso, 2012).

© The Author(s) 2019
J. M. Robles-Morales and A. M. Córdoba-Hernández,
Digital Political Participation, Social Networks and Big Data,
https://doi.org/10.1007/978-3-030-27757-4_7

Garza,[2] an activist who was watching the news in a bar in Oakland, California, expressed her indignation with a post on her Facebook account which said: "*Black people. I love you. I love us. Our lives matter*" (Brown, 2015).

A friend of hers, Patrisse M. Cullors,[3] saw the post and commented on it the same day, using the hashtag #BlackLivesMatter (hereafter #BLM), based on the last three words of Garza's post. Two days later, Cullors and Garza spoke for the first time about a shared project #BLM, which they hoped would bring many people together, revealing what it meant to be "black" in the United States (Brown, 2015).

A short time later, Opal Tometi[4] called both activists to propose the creation of a platform for the project, completing the trio of founders of #BLM (Craven, 2015). From the beginning, Tometi was in charge of designing the communications plan and social media strategies to bring the largest possible number of people together.

Further cases of police abuse against black people continued to appear. The video of the death of Eric Garner, on 17 July 2014, after an electric shock administered by a policeman, Daniel Pantaleo, in New York, while he was being arrested for the alleged sale of contraband cigarettes, spread quickly across social media. On this occasion, the courts of New York decided not to prosecute Pantaleo, which led to huge protests in New York, Louisiana, Ohio, California and Massachusetts, among other places (Goodman, 2014).

The incident that led to the greatest public outcry occurred three weeks later, on 9 August 2014, when Darren Wilson, a white police officer in Ferguson, Missouri, shot Michael Brown, an 18-year-old African American who broke into a shop with some friends to steal cigarettes.[5] Despite the

---

[2] Besides being one of the young co-founders of the Black Lives Matter movement, Garza is the director of special projects for the *National Domestic Workers Alliance* (NDWA) and is a columnist for newspaper such as: *The Guardian, The Nation, The Feminist Wire* and *Huffington Post* (Brydum, 2015).

[3] A young activist who had founded *Dignity and Power Now* in 2012, an NGO based in Los Angeles that fought for the rights and dignity of incarcerated people, their families and communities (Brydum, 2015).

[4] Activist and writer, she was also the executive director of the *Black Alliance for Just Immigration*, an organization that defends the rights of black and Latin American immigrants in the United States (Dalton, 2015).

[5] According to the report issued by the Justice Department of the United States, Michael Brown and his friends stole some cigarettes from a shop without any kind of weapons. Wilson received a call about a "robbery under way" went to the site and blocked access with his

evidence, Wilson was absolved and acquitted of murder, as Zimmerman had been a year previously. The atmosphere was already highly charged, and with #BLM expanding across social media, public demonstrations against the decision and in defence of the civil rights of the black population appeared immediately and escalated into several weeks of violence and tension between the demonstrators and the police.

Although #BLM was first coined in July 2013, it was not a popular hashtag in its first year. Statistics show that it was only used in 48 tweets prior to June 2014. With the death of Garner, in July, it spread a little more and was used in 398 Twitter posts. It was not until the events in Ferguson that it became truly viral, being used 52,288 times in August 2014 and emerging as the most symbolic hashtag of the Missouri protests (Freelon, McIlwain & Clark, 2016) and one of the most viral in 2014 on the network.

After the success of the mobilizations in August, Garza, Cullors and Tometi, decided to give Black Lives Matter a more consistent structure and turned it into an online social movement, supported by chapters in different cities which had to be approved a central authority and linked to official communications channels.[6]

Since then, many citizens have wanted to take part in the mobilizations organized by BLM and which, in one form or another, express the social discontent caused by discrimination and racial inequality in the United States. People get involved by taking part in events that are announced online, or by contacting their nearest chapter. The activities are financed through the sale of BLM products and donations, such as those by Jay Z and Beyoncé, who have paid the bail demanded for activists arrested in the demonstrations (Pearson, 2015).

In view of the scope that the hashtag has achieved and the organization and structure of the movement, we can see there is a clear distinction between #BlackLivesMatter and its repercussions in public debate and the Black Lives Matter movement, which has a structure that coordinates its online and off-line events and organized more than 1200 public demonstrations of support across the whole United States in its first two years (Cañuelas, 2016).

---

vehicle. Wilson declared that Brown tried to reach the vehicle and he had no option but to use his weapon (Department of Justice, 2015).

[6] In August 2017, according to the BLM web page, there were 38 *chapters* in all, distributed across different cities of the United States and Toronto (Canada).

If we compare BLM with the civil rights movement, we can see that despite having a lot in common, there are considerable differences in their ideological foundations. The civil rights movement under Martin Luther King Jr. combined three key factors: the presence of a charismatic leader, activism with strong roots in Christianity and the predominance of male roles over female ones. These are not present in BLM: the religious convictions of the participants have no bearing on the ideology behind the movement; the leaders who speak for the different chapters can be male, female or transgender, and there is no charismatic leader like there was in the past.

It should also be noted that the current trend for inclusion and solidarity between social movements means that the struggles of some movements are adopted as part of the struggles of others. For example, one of the principles of BLM is to include the fight for LGTB rights, and vice versa. It is therefore a network of complex movements that support each other to enable each other to survive and extend their scope at the same time as they demand equality.

While the BLM movement acquired consistency with its own structures, the hashtag continued to gain ground thanks to support by many celebrities who adopted #BLM on their social media accounts. This includes tennis player Serena Williams, Democratic Party politicians, singers like Katy Perry, Justin Timberlake, Kanye West and even Prince, who referred to the movement in the 2015 Grammy awards of 2015, shortly before his death. Kim Kardashian for her part, has spoken extensively on the subject, posting an essay on her website in which she praises the principles of Black Lives Matter.

Since the events in Ferguson in 2014, #BLM has raised public awareness in North America and spurred the desire to press for reforms that guarantee equal rights for African Americans among the white as well as the black population, as shown by recent surveys (Menasce & Livingston, 2016).

The hashtag is still alive, although dormant, on social networks, and reappears every time there is an act of violence against the African American community, triggering increased distribution, visibility and international support.

The clearest indication of support for the principles behind #BLM is its adoption by politicians. This was the case in the statements made by Barack Obama on 7 July 2016 in response to the death of Alton Sterling in Louisiana and Philando Castile in Minnesota. He proposed far-reaching reform of the security services in the United States and said that this was not just a problem for the black population but for America, and it should be a concern for the whole country.

Police aggression against black citizens may continue, but #BLM managed to articulate a sense of public indignation that put the issue at the centre of attention in a way never seen before. Videos, images and reports of violent encounters between the police and African Americans are a permanent feature in the media and on social networks. What Garza, Tometi and Cullors set up in 2013 has become a topic of public debate, and #BLM has been adopted by different citizens without them necessarily forming part of the Black Lives Matter Movement. Its opposition has eventually emerged in another concept coined by cyber-language: #AllLivesMatter, the motto of those who criticize the movement.

7.2   #BLM Symbolic Inclusiveness

Once we have understood the evolution of #BLM we can go on to analyse how it has disintermediated the message on social media. According to the concepts we looked at in Chapter 2, we can classify it as a case of connective action in which we can clearly see both elements that form part of these processes (Bennett & Segerberg, 2013):

1. *Symbolic inclusiveness*: political content and messages that can be easily personalized and shared by many people who identify with a common problem.
2. *Technological openness*: connective actions make use of the range of possibilities offered by technology to share these inclusive messages on different platforms and in different formats.

Both converge in a phenomenon of political personalization through digital media. Today's citizens expect greater flexibility in joining causes, ideas and organizations, with individual actions and through social media. This type of mobilization does not rely on the financial resources of the organizing body, nor strong collective identities, but rather acts of personal expression and recognition available to each individual in the digital sphere (Villanueva-Mansilla, 2016).

Returning to the first element, symbolic inclusiveness only takes place when the message can be adapted and personalized in different contexts.

This is precisely where Occupy Wall Street succeeded, with its famous slogan: "*we are the 99%*". This is a flexible, broad and malleable frame which many people could identify with as part of this 99% of the population, as opposed to the elite who made up 1% of North American society, according to this group.

Something similar happened with #BLM, which was modified by the users according to their personal experiences, which are after all the main drivers of the concerns on our individual agendas and from where we find our points of contact with other citizens (Sánchez & Magallón, 2016).

One strong indication of this personal aspect can be seen in the Twitter users who adopted #BLM as their profile name. In December 2015, a year and a half after the Ferguson demonstrations, there were 94 people on Twitter with this username. We can consider this an example of symbolic inclusiveness, because only 5 of these accounts were created after the hashtag appeared in the network, while the others changed their name after it appeared.

What is interesting about these 94 accounts is that only three of them are for groups or organizations,[7] but they are not verified accounts or related to any official account of the Movement. The other 91 belong to individuals who identified with the cause, giving it a personal frame. Logically, most of the users are located in the United States or Canada, where the chapters of the Movement are, but there are others from other countries, including: Japan, France and the United Kingdom. This shows how global the hashtag became and how it was adapted to different contexts.

A similar thing happens with the user profiles. Although this data is not easily available, because few accounts include this information, the description offered by the users is striking because they are very diverse: feminists, homosexuals, students, black people, artists, poets and musicians, among others.

Once we have found these accounts, we want to see their activity on Twitter with #BLM: how often they use it and the topics the tag is associated with, to compare them to the movement's official account. In one year, these 94 accounts posted 544 tweets with the hashtag, but they were generally infrequent. Curiously, 44 of these accounts made no tweets with the hashtag, despite using BLM as their username. They would appear to

---

[7] @BLM_INC; @blmpasadena and @BLMFLORIDA.

Table 7.1 Frequency of the appearance of other hashtags in tweets on the official account and the other accounts

Official account		Various accounts	
Hashtag	*Frequency*	*Hashtag*	*Frequency*
#EricGarner	24	**#Ferguson**	35
#ICantBreathe	21	#SandraBland	21
#twocc	15	#SayHerName	19
#ShutItDown	14	#UniteBlue	18
#fergusonfridays	13	**#EricGarner**	17
#Ferguson	11	**#ICantBreathe**	16
#NYC2Palestine	11	#AllLivesMatter	16
#BlackTransRevolution	10	#PoliceBrutality	13
#cantkillafrica	10	#BlackTwitter	13
#ThisStopsToday	9	#shutdownOPD	12

Source Twitonomy. Own research

be people who identify with the cause but do not publish content related to it.

As regards the topics associated with the tag, there is yet more variety. When comparing all the tweets from these accounts that used #BLM against the official account of the movement over one year, there were 556 different hashtags. If we take the 10 most recurring tags of each one, we can see there is disparity in the discourse, because only three of them are the same. #EricGarner, #ICantBreathe and #Ferguson (Table 7.1).[8]

Some tags like #Twocc and #BlackTransRevolution appear in the official discourse of the movement as one of its ideological motives: in defence of the Trans Women of Colour Collective, but do not appear on the other accounts. In contrast, #AllLivesMatter, which is one of the most recurring hashtags in the tweets from these users is not used once by the official account, because it is seen as a hashtag that perpetuates white domination by asserting that all people are treated equally (Black Lives Matter, n.d.).

To end this section, we would like to classify the tags by topics into three groups. The first group would include the hashtags that mention the victims or recall the situations in which they died; the second would include

[8]The first two were in memory of Eric Garner, murdered in New York on 17 July 2014 and his last words before dying from the electric shock applied by a police officer. The third is the symbol of the Ferguson protests in August 2014.

Table 7.2   Examples of other hashtags that use the words: black, lives or matter

Black	Matter	Black + Matter
#BlackAmerica	#AllLivesMatter	#BlackTransLivesMatter
#BlackBrunch	#NativeLivesMatter	#BlackGirlsMatter
#BlackChurchesBurning	#MySonsLifeMatter	#BlackKidsMatter
#BlackFutureMonth	#BlueLivesMatter	#BlackMenMatter
#BlackoutBlackFriday	#TransLivesMatter	#BlackLivesMatterLansing

Source Twitonomy. Own research

the hashtags that reference the places where the protests were held and the third would reference all the tags in which the message of #BLM is transformed in some way, containing the words, *black, lives* or *matter,* but combined with other words, such as: #3rdWorld4BlackPower, #BlackDollarsMatter, #BlackSolidarityMatters, #BlackVote4Bernie, #BlackXmas, etc. (Table 7.2).

In general, the word *black* was related with events or actions by Black Lives Matter, such as the *blackout* campaign or the Black Friday boycott. Hashtags with the word *matter* were usually related to other groups, not necessarily black, such as Native Americans. Finally, when the hashtags contained both words, the topics focused on specific groups of citizens: transsexuals, women and children, inter alia.

7.3   TECHNOLOGICAL APERTURE

The second element proposed by Bennett and Segerberg (2013) as part of connective actions in technological aperture, in which citizens can take advantage of the structure of the Internet to spread a message and extend its scope. This is similar to what Jenkins, Ford, and Green (2012) called *spreadable media*: media that can spread over social surfaces like a paste. The content is not spread from specific central points but slowly seeps through society and can be used to extend that media's influence across vast surfaces.

On the one hand, technological aperture can be followed through the "layers" of the Internet its content spreads over, meaning all the digital tools where the message arrived by itself without any official channels of the movement, such as the use of links to reach other web pages or sites.

We can also consider the resources that the group has available to spread its message as technological aperture. As we have seen, BLM has a series of "official communication channels" on the web which are designed specifically for information, to make content visible and accessible.

If we go on the BLM website, the users can be redirected to the 4 official accounts on: Facebook, Twitter, Tumblr and Instagram, which bring together nearly 650,000 followers and fans in total. However, technological aperture also includes the number of unofficial accounts and pages where the cause is also shared, and which cannot be controlled or manipulated by the BLM movement.

We have already seen the 94 Twitter accounts which adopted the name Black Lives Matter and which are not verified by Twitter as official accounts of the movement. The same thing happens on other social networks. For example, the movement had an official Facebook page with 222,542 fans at the end of 2016. There were, however, another 72 accounts with this name, of which only 19 are verified as the accounts of different chapters, and they add some 593,754 more fans.

Thanks to the connections between platforms and technological resources, activists and citizens that identify with a cause have been able to share text, images, videos, posts, audios, etc. on their personal devices, keeping the message of police brutality against black people alive, making the personalization of the message more common through digital connections with friends, followers and fans, depending on the different formats of the networks (Bennett & Segerberg, 2013).

7.4 From Ideology to Identity and from Identity to Individuality

In the sixties and seventies, a new type of social movement appeared, first in the United States and then in other parts of the world. Unlike traditional social movements, such as the labour movement, the bond of union among its members was no longer a matter of ideology or class. Classical movements such as the labour movement or suffragism were characterized by bringing together people who shared problems of their class, gender or race. Among these, naturally, the conditions of work and social exclusion generated by these conditions or the search for political rights that, because they belonged to a certain gender or race, were vetoed.

However, and largely because of the development and improvement of living conditions and the extension of social and political rights, the social movements of the 1960s and 1970s have a peculiar characteristic; they focus on problems that have to do with identity and with aspects of private life. They are, from this point of view, much more transversal. The fight for the rights of homosexuals and lesbians is a recurring example of this type of movement. It does not matter the social class or the colour of the skin. This type of movement pursues the recognition of the rights of people who have a sexual option different from what the society of the seventies considered "normal". It is, precisely, the fight for these rights, one of the key generators of identity. Collective action and belonging to this type of movement generates identity and this identity generates collective action and organizations of social movements. The activists of these new social movements do not have a unique identity, class identity, but a multiple identity that, to a large extent, comes from the social movements in which they participate.

However, both in traditional movements, motivated by class, race or gender conditions, and in movements motivated by identity issues, organization was a key issue. In the case of classical movements, the organization allowed, for example, the organization of collective action. In the case of the new social movements, the organization was the sphere of socialization and construction of the values that define the collective identity.

Processes such as those described herein the context the analysis of Black Lives Matter, represent a substantive change with respect to classical and new movements. This change has to do with individuality. It is another step that has taken us from the plurality of identities to individual identities. The main effect of social networks like Twitter or Facebook is that they generate the possibility of adding you to particular and concrete causes in a personalized way. It is not necessary to know much about the cause, it is not necessary to belong to an organization, it is not necessary to dedicate much effort and time of your day. It is enough that the cause seems just/interesting to you.

It is a process of disintermediation because traditional organizations, in this case social movements, no longer function as the sources of resources for collective action (generators of messages and a discourse that explains reality) or as the seed for definition of identity (the values that define us as members of a collective). The information and messages that flow through the network (sent by other people) are adopted according to the personal disposition of the one who is exposed to them. Naturally, the preferences and attitudes that lead a person to support #BLM are socially constructed. That is, they depend, in one way or another, on the relationship of people with institutions such as the family or social and generational groups. However, the disintermediation of the messages, their growing independence from strongly structured organizations allows activism on demand.

This digital activism is key to our democracies because, thanks to it, the message of #BLM has been able to go very far. The disintermediation of messages plays in favour of their virality and the disposition of many people outside the germ of this process, to join the cause. However, there is no doubt that the digital context that disintermediates messages is based on weak commitment bonds. Not only because, as has been said repeatedly, the democracy of slacktivism is a poor democracy, but because the disintermediation of messages, as we have seen here, filters collective values through the sieve of individual tastes. There is no use in digital activism, but a self that supports causes without a deep compromise with them. We believe that a society that is structured on the basis of activist egocentrism is a society that moves away from those values that are represented by the men and women who dedicated their lives, their days and their nights, to the achievement of the goals they pursued. We have the feeling that the individualization of digital communication opens the door to very serious problems that we will analyse in Chapter 10 this book.

References

Bennett, W. L., & Segerberg, A. (2013). *The logic of connective action: Digital media and the personalization of contentious politics.* New York, NY: Cambridge University Press.

Black Lives Matter. (n.d.). *Guiding principles.* Retrieved from http://blacklivesmatter.com/guiding-principles/.

Brown, J. (2015, August 7). *One year after Michael Brown: How A hashtag changed social protest.* Retrieved from http://www.vocativ.com/218365/michael-brown-and-black-lives-matter/.

Brydum, S. (2015, December 8). *Patrisse Cullors knows that we are not separate.* Retrieved from http://www.advocate.com/40-under-40/2015/12/08/patrisse-cullors-knows-we-are-not-separate.

Cañuelas, L. (2016). Comunicando la violencia: el caso de Black Lives Matter. In *IV Conferencia Conjunciones Complejas: Encuentro transdisciplinario para el estudiode la violencia.* San Juan, Puerto Rico.

Craven, J. (2015, September 30). *Black Lives Matter co-founder reflects on the origins of the movement.* Retrieved from http://www.huffingtonpost.com/entry/black-lives-matter-opal-tometi_us_560c1c59e4b0768127003227.

Dalton, D. (2015, May 4). *The three women behind the Black Lives Matter Movement*. Retrieved from http://madamenoire.com/528287/the-three-women-behind-the-black-lives-matter-movement/.

Department of Justice. (2015, March 4). *Report regarding the criminal investigation into the shooting death of Michael Brown by Ferguson, Missouri police officer Darren Wilson*. Retrieved from http://www.justice.gov/sites/default/files/opa/press-releases/attachments/2015/.

Freelon, D. G., McIlwain, C. D., & Clark, M. D. (2016). *Beyond the hashtags:# Ferguson,# Blacklivesmatter, and the online struggle for offline justice*. Washington, DC: Center for Media & Social Impact.

Goodman, J. D. (2014, August 4). Difficult decisions ahead in responding to police Chokehold homicide. *The New York Times*. Retrieved from http://www.nytimes.com/2014/08/05/nyregion/after-eric-garner-chokehold-prosecuting-police-is-an-option.html?smid=pl-share&_r=0.

Gutman, M., & Tienabeso, S. (2012, March 2). Trayvon Martin shooter told cops teenager went for his gun. *ABC News*. Retrieved from http://abcnews.go.com/US/trayvon-martin-shooter-teenager-gun/story?id=16000239.

Jenkins, H., Ford, S., & Green, J. (2012). *Spreadable media: Creating value and meaning in a networked culture*. New York: New York University Press.

Menasce, J., & Livingston, G. (2016, July 8). *How Americans view the Black Lives Matter movement*. Facttank, News in the Numbers. https://www.pewresearch.org/fact-tank/2016/07/08/how-americans-view-the-black-lives-matter-movement/.

Pearson, M. (2015, May 9). Jay Z posted bail for protesters, writer says. *CNN.com*. Retrieved from http://edition.cnn.com/2015/05/19/entertainment/feat-jay-z-protesters-bail/.

Sánchez, J. M., & Magallón, R. (2016). Estrategias de organización y acción política digital. *Revista de la Asociación Española de Investigación de la Comunicación, 3*, 9–16.

Villanueva-Mansilla, E. (2016). Acción conectiva, acción colectiva y medios digitales: posibilidades para la comunicación política en los tiempos de Internet. *Contratexto, 24*, 57–76.

CHAPTER 8

The Disintermediation of the Space: The Case of #BringBackOurGirls

8.1 The Kidnapping of the Chibok Girls Takes Over the Internet

We have seen a marked increase in recent years of citizens making use of social networks to take part in global campaigns. We only need to think of the video KONY 2012, by the Invisible Children organization or the #IceBucketChallenge solidarity campaign to raise funds, which achieved millions of views and beat all previous records on YouTube. Even so, some authors have been deeply critical of this type of participation, because they consider that this type of mobilization on social media should not be seen as true participation, but a simple exchange of information without real, personal commitment to the causes.

In this section, we shall be using as an example the repercussions of the hashtag #BringBackOurGirls, which was a protest against the kidnapping of 276 girls by Boko Haram in Nigeria, and which was one of the most viral tags of 2014. The underlying question is whether this type of global mobilization, which critics have branded "slacktivism" (Christensen, 2011; Glenn, 2015; Morozov, 2009), a portmanteau of slacker and activism, can have any effect on the real world and overcome the spatial barriers that other political actions face.

Before starting our analysis, let's recall the context of the case. Boko Haram was founded in the north of Nigeria in 2002. Its literal meaning in Hausa is: *Western education is sinful*. From the start, this *jihadi* group

© The Author(s) 2019

J. M. Robles-Morales and A. M. Córdoba-Hernández,
Digital Political Participation, Social Networks and Big Data,
https://doi.org/10.1007/978-3-030-27757-4_8

considered itself to be the most loyal followers of the Taliban in Africa, and their intention was to impose Sharia law in Nigeria through sectarian warfare.

The terrorist group only raised concerns in the West when it claimed responsibility for a series of attacks on Christian communities over the Christmas period in 2011. The General Secretary of the United Nations, Ban Ki-Moon, immediately expressed his condolences and sympathy for the families of the victims, along with the White House, which spoke out against the attacks (*El País*, 2011).

Despite the other terrorist attacks carried out by the group, what really shocked the whole world was the kidnapping of 276 girls aged between 15 and 18 from a secondary school in Chibok, Borno (Nigeria) on 14 April 2014.

A few days later, 53 of the girls managed to escape their captors, but the authorities had no idea where the other 223 in the hands of Boko Haram might be. After the kidnapping, Aboubakar Shekau, the main leader of the terrorists, issued a video in which he said that the girls were in his power and that he would sell them as slaves.

To express their rejection, families of the girls began to use the message "Bring Back Our Girls" in several demonstrations that were joined by other organizations in a campaign to pressure the Government of the President, Goodluck Johnson, into taking more determined steps in the search for the minors. However, it was a Nigerian lawyer who used the hashtag #BringBackOurGirls (hereafter #BBOG) for the first time on 23 April 2014 when commenting on a speech at a UNESCO event (Lyons, Robinson, & Chorley, 2014).

In just two weeks, this hashtag was being used around the world (Bajo, 2014; Collins, 2014). The most popular action of the campaign, in which many celebrities took part, was to publish a photo in social media accounts such as Facebook, Twitter or Instagram, holding a card which said #BBOG, as a kind of "war cry" in favour of the girls.

Celebrities were quick to join in, on 7 May, Malala Yousafzai, Michelle Obama and Sean Penn joined the campaign with the hashtag #BringBack-OurGirls and the sentence "Real Men Don't Buy Girls", posting photos of themselves with these messages on their Twitter accounts (*El País*, 2014). Three days later, the first lady of the United States devoted a whole television programme to supporting the girls and their families (Fernández, 2014).

The viral impact of the hashtag on the networks meant that expressions of support from around the world multiplied and images appeared of protests in different cities in the United States and Europe. Artists, politicians and celebrities joined in the demonstrations carrying placards in well-known locations such as the Eiffel Tower or the White House.

The impact of the campaign around the world prompted Aboubakar Shekau to publish a second video on 12 May 2014 in which he showed hundreds of girls wearing veils and proposed to exchange them for Boko Haram prisoners being held by the Nigerian Government (Naranjo, 2014).

This same public pressure may have led the governments of several countries to express their concern and to contact the Nigerian government to offer support in the search for the girls. On 24 May, following a meeting in Paris between the President of Nigeria and the foreign ministers of the United States, Israel. France and the United Kingdom, it was agreed that there would be no negotiation with the terrorists, but a coordinated military rescue operation.

Several governments, including the United States and United Kingdom offered to send military expertise and intelligence to the area to collaborate with the local authorities in the operation. Other countries of the European Union offered to support the search and rescue of the captives, as did Canada. China stated that it would share important information with the Nigerian Government and even the Foreign Minister of Iran labelled Boko Haram as "Takfires" (false Muslims) and offered help to resolve the crisis.

At the same time, NGOs and international feminist organizations such as FEMNET and Igualdad Ya concentrated their efforts on organizing the global #BringBackOurGirls campaign, gathering support to call for the return of the girls.

8.2   Leaping the Barriers of Space with "Slacktivism"

The global campaign that #BBOG turned into can be seen as a phenomenon of digital activism that has generated its own form of political intervention. First of all, it confirms that the notable reduction in the cost of participation through the Internet motivates citizen mobilization. The means offered by social networks enables many individuals to make small contributions on their personal accounts that converge into a major mobilization (Butler, 2011; García & del Hoyo, 2013).

Alongside these changes in the resources and costs required for participation, the forms of mobilization have also been changed. These actions reveal a new type of activist, the "click-activist", whose political identification is flexible, requiring less personal commitment and with fewer ideological tensions (Butler, 2011; Henríquez, 2011; Resina de la fuente, 2010). These campaigns have been denigrated as mere slacktivism, a type of social media activism that entails little cost or risk and seeks only to provide emotional satisfaction and show the sensitivity of the people who share the activity, but which has no real effect on politics or society.

What type of actions does slacktivism include? All initiatives in which anyone with an Internet connection can take part in: "liking" to show appreciation in platforms like Facebook or Instagram; making hashtags and content on Twitter go viral; signing online petitions; resending messages, videos and other content through social network accounts; donating small amounts of money; changing the status of personal profiles; creating causes in Facebook; posting photos and selfies to show support for a campaign, etc. (Christensen, 2011).

Of course, not everything that circulates on social media should be considered slacktivism, but only those actions that are clearly trying to influence political decisions. In this case, we should ask ourselves: What was #BBOG's greatest achievement? Visibility, taking the cause outside the national context, enlisting international support and being able to apply pressure on the Nigerian government to make more efforts to find the girls.

Little by little, the hashtag first used by the Nigerian lawyer turned into a truly global campaign to achieve the visibility that was needed to prevent the kidnapping of the girls from being merely an isolated case in an African state, by making the rescue of the girls an international cause. To achieve this, the #BBOG campaign contained two vital elements: the viral growth of the tag in the first two weeks and the influence of public figures who placed the issue firmly on the international agenda. The goal was always the same, to achieve the "boomerang effect" (Vegh, 2003) in which a cause reaches out beyond the barriers of national frontiers to return invigorated by worldwide support.

As regards the first of these elements, we are aware that "virality" is not an accepted dictionary term, but technical jargon to describe a phenomenon in which some content, usually of a highly emotional nature, is spread massively across different social and traditional media (Berger & Milkman, 2012). In this case, the viral nature is essential because it kept the hashtag alive and quickly aroused interest among users.

The second element, however, which we have termed "influence", in line with the title of "influencer" that has become closely linked with social media (Bakshy, Hofman, Mason, & Watts, 2011), refers to the effect on a campaign like this one of the adoption of the hashtag by influential accounts and users, and their use of it in their personal content. We could say that the success of a campaign may well depend on the ability of its messages and hashtags to enrol influencers.

These two categories of virality and influence can tell us how it was that the spread of the #BBOG tag achieved such visibility through a phenomenon of global mobilization and the characteristics of the most influential users.

8.3   Virality and Influencers as Drivers of Visibility

To fully examine the achievement of #BBOG, we must look at the behaviour of the hashtag on Twitter between April and December 2014. We want to see its viral expansion across the Internet and the influencers who boosted the hashtag over this year.

There are many commercial online APIs that we can use to monitor users and hashtags in Twitter. In this case we used six of them: *Topsy, TalkWalker, Twitonomy, TweetArchivist, CartoDB* and *Hastagify.me*; we applied them at different periods of the year to analyse the behaviour of #BBOG and its influencers in 2014.

To discover how viral the hashtag was, we can use interactive analytical tools. According to the data collected by *Topsy*, in just two weeks, the hashtag became a trending topic in Nigeria and reached 1,334,864 posts worldwide, with interesting centres of attention outside the African country, in which the movement in the United States and Europe was higher than that of Western African countries (Neubauer, 2014). In this sense, the Nigerian cause grew in a similar way to the protest #StopKonny in 2012, in the sense that it appeared in Africa but quickly became more popular in the United States and Europe, spreading from there to other continents.

At the end of 2014, several girls remained in the hands of Boko Haram and #BBOG was still active on Twitter. At that time, we measured the virality of the data in real time over one week, using the API TalkWalker to see where the posts came from and in which language they were written.

Eight months later, the poles of virality were still the same. The hashtag showed greater activity in the United States over this week with 4600 tweets. What was different this time was that Nigeria and its neighbouring countries, with 3800 tweets, pushed the European nations into third place with 1800 tweets (Image 8.1).

The second element that we wanted to measure in this case was the role of influencers in making the campaign more visible. If the actions of the #BBOG campaign are understood as part of a *slacktivist* strategy, the increased virality of the hashtag should be seen as a fundamental part of this, in which users spread the tag either through messages that are sent out to many people (number of followers) or through the frequency of reposts (retweets using the hashtag).

To measure this, we used the *Hashtagify.me* platform to discover that the six most influential users of the hashtag in 2014 were, in order: the actress Emma Watson, the presenter Ellen DeGeneres, the singers Katy Perry and Chris Brown; UNICEF and CNN.

The idea was to see which accounts raised #BBOG to its highest level of popularity through the number of times it subsequently appeared on the timeline (TL) of other users, despite them not using the hashtag frequently in their own posts.

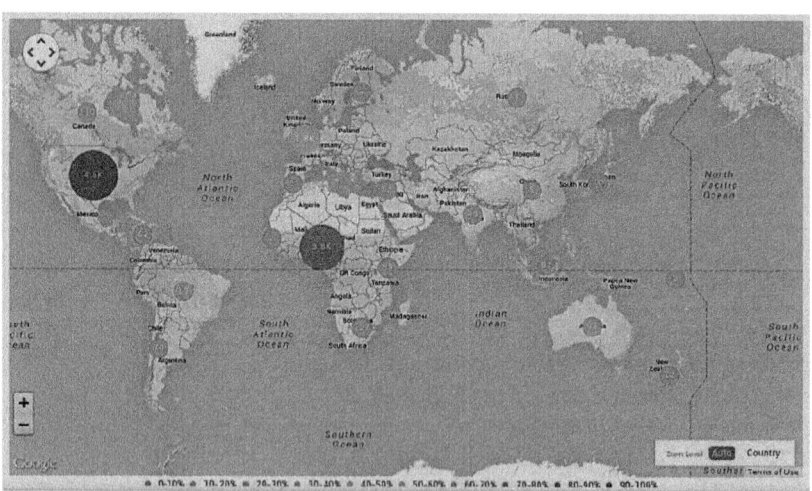

Image 8.1 Map of viral growth of #BringBackOurGirls (4–11 December 2014) (*Source* Talkwalker)

Let's stop to look at the case of Emma Watson, the most influential account for the campaign over the year. The actress only used the hashtag twice, on 9 and 10 May 2014, and in one of these she appeared with one of the familiar strategies of the Nigerian campaign, holding a card which had the hashtag written on it. Adding together the numbers of both tweets, the account achieved 62,000 retweets and 93,000 favourites in two days. Despite the fact that she did not say anything more on the issue, the celebrity considerably increased the popularity of the action (Image 8.2).

It may seem remarkable that an account achieves such visibility for the hashtag, but it can be explained by user behaviour on Twitter. As we would expect, the celebrity accounts among the six influencers are not particularly active on the network. For example, at the time of the analysis, Emma Watson had only published 817 tweets in four and a half years but had the highest average number of RTs for each post (5115.17). The account of a news outlet as well-known as CNN, with 52,822 tweets in total, only has an average of 450.07 RT for each tweet (Graph 8.1).

As we did when researching virality, in December of the same year we wanted to know which influencers were using the hashtag eight months later, when the topic was no longer a central point on the agenda and public debate, but while the girls were still in the hands of Boko Haram.

The findings drawn from the data provided by Topsy were interesting because the hashtag continued to be active in November and December of that year, peaking with a Tweet on the account @bbcbreaking on 13 November that appeared 2852 times on the timelines of other users.

We also wanted to analyse the profiles of the six most influential users of this month according to *Hashtagify.me*, as we had in the first two weeks after the appearance of the tag. On this occasion we saw that the accounts that were doing most to make the hashtag visible were no longer public figures, the media or organizations, but mostly citizens, activists and Nigerian politicians.

The trend was different, and unlike April and May, these were little-known accounts, but with a much higher average of tweets per day, as many as 500 or 780 every day, but which did not achieve the same RT levels as the celebrities earlier in the year.

We can therefore conclude that #BBOG is an example of slacktivism. Between April and May 2014, the hashtag was at the top of the media agenda and became a trending topic on social media. Many celebrities and politicians posted photos on their accounts posing with cards with the tag written on them. Some never used the tag and probably are not aware of

Emma Watson ✔
@EmmaWatson

#BringBackOurGirls

※ BRING
BACK OUR
GIRLS

6:41 - 10 may. 2014

40.633 Retweets **67.534** Me gusta

Image 8.2 Message by @EmWatson in which she uses the hashtag (10 May 2014) (*Source* Twitter)

what happened to the girls but could not resist joining in on the worldwide protest that was happening. Nevertheless, and probably unwittingly, their contribution helped to spread the news of a cause that might otherwise have gone unnoticed.

These slacktivists are part of a reactive audience who occasionally take part in some processes, directly and independently, through the potential offered by the Internet, but go no further than this. The scope of

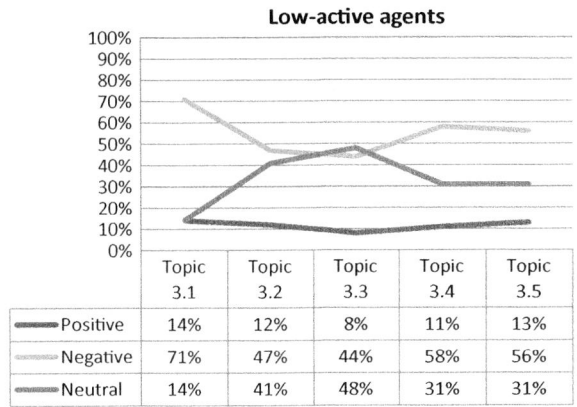

Low-active agents

	Topic 3.1	Topic 3.2	Topic 3.3	Topic 3.4	Topic 3.5
Positive	14%	12%	8%	11%	13%
Negative	71%	47%	44%	58%	56%
Neutral	14%	41%	48%	31%	31%

Graph 8.1   Number of tweets published by the most influential users and the average retweets achieved by each (15 December 2014) (*Source* Own research, based on Twitonomy)

this resource, however, is measured by other standards. These actions do not strive for conventional political commitment, their effectiveness lies in being a driver of opinion and visibility in order to generate awareness and international solidarity. In the case of the kidnapped Nigerian girls, the wave of public opinion that #BBOG unleashed on the Internet drew the attention of countries like: France, the United States and United Kingdom, who put pressure on and backed the President of Nigeria in stepping up the search for the girls.

One interesting aspect of the case in the curve shown by the hashtag. It initially came from a Nigerian lawyer with no online influence, but it gained mass popularity after it was picked up by the media and celebrities in the intermediating sphere in the United States and Europe, which was sufficient to promote the cause. This was when the slacktivism kicked in, as the result of the worldwide concern. Eight months later, the profiles of the most influential users employing the hashtag in Twitter had changed to a more local and political type: Nigerian citizens, activists and politicians. These profiles are more committed to the cause and track the situation of the girls every day.

We believe that, thanks to social networks, geographical barriers (spaces) are not as strong an obstacle as that which existed only a decade and a half ago. Without the social networks and without their capacity to break the physical spaces, perhaps the global public opinion would never have attended to the situation of this group of girls. The media rupture of geographic barriers to generate attention on a demand for support or help is not something new. Our childhood was marked by the terrible eruption of El Nevado del Ruiz volcano, in 1985, and the terrible consequences it had for the people of the surrounding towns, especially for the town of Armero. In this and other events, spatial barriers gave way and generated solidarity and support for the victims of people from all parts of the world.

However, this rupture of space and time depended on the mediating action of the mass media. La desintermediación de espacios se entiende como la capacidad de nuevos agentes, nuevos mediadores, para romper con los espacios geográficos y buscar apoyo para causas que, quizás, no serían públicas. This, from our point of view, is a profoundly positive result of communication mediated by social networks.

The case that we have presented here is an extreme case. It is about the lives of people who, in addition, are minors. Nothing can obscure the fact that it is a legitimate and universal cause. However, the disintermediation of spaces has aspects that are somewhat less clear. We refer to how causes are judged and how solutions are defined. There is no single possible solution for issues that, like those that interest us, generate public debate. Thus, we coexist with situations that, for some are acceptable and normal, while, for others, they are intolerable or negative. Suffice it to recall, for example, the profound debate generated in France during the first years of this century about the use of the characteristic veil of Muslim women in public spaces.

However, the disintermediation of the spaces seems to suppose that, once a topic has gone viral, it is already on the way to a solution. Obviously, humanitarian causes and those that affect the lives of people are easily definable within the framework of human rights. However, what happens to cause that, like that of ethnic minorities or those that are strongly opposed to economic power, are more linked to ideological or political issues. In these cases, the capacity to become transversal or viral descends and the spaces where the people to whom these problems live, remain obscured. The disintermediation of the spaces is not innocent. Some spaces are lit, and others are kept in darkness not for reasons of chance, but for the capacity of these spaces to attract the attention of viral public opinion. This public opinion is more willing to retweet clear issues, causes that can be expressed with a for or against, that those that require a more elaborate reflection and argumentation.

The other question here, are the solutions. We live, as Morozov (2014) says at the time of technological "solutionism". This concept refers to the tendency to consider that every problem has a unique and univocal solution. Said solution is, most of the time, a technical one. A solution that can be modeled and applied to all those problems that are similar. This form of global technological solution ignores the specificities and the specific characteristics of the problems. That is, to its local nature. The disintermediation of spaces makes possible the propagation of technological solutionism in that it proposes solutions for problems that affect others (in many cases people who are in faraway countries) through technical principles that are typical of rich countries.

This study does not aspire to align itself with the idealistic view that social networks are transforming our social and political structures. Nor does it adopt the opposing, pessimistic viewpoint of considering that no change has occurred. The only thing it seeks to show is the possibilities of actions like those we mentioned earlier. While for a mobilization to be effective it has to be accompanied by conventional strategies for participation, this does not mean that slacktivism is a resource that makes no contribution to raising awareness on and increasing the visibility of a cause.

#BBOG is a typical case of slacktivism but, instead of condemning these campaigns because they do not lead to effective political action, this work seeks to find a proper place for them and their specific effectiveness as resources for mobilization.

References

Bajo, C. (2014, May 14). El 'hashtag' del secuestro. El secuestro del 'hashtag'. *El País*. Retrieved from: http://elpais.com/elpais/2014/05/13/planeta_futuro/1399988511_367696.html.

Bakshy, E., Hofman, J. M., Mason, W. A., & Watts, D. J. (2011). Everyone's an influencer: Quantifying influence on Twitter. In *Proceedings of the Fourth ACM International Conference on Web Search and Data Mining*, pp. 65–74.

Berger, J., & Milkman, K. (2012). What makes online content viral? *Journal of Marketing Research, 49*(2), 192–205.

Butler, M. (2011). *Clicktivism, slacktivism, or 'real' activism, cultural codes of American activism in the Internet era* (Master's thesis). University of Colorado.

Christensen, H. (2011). Political activities on the Internet: Slacktivism or political participation by other means. *First Monday, 16*(3). Retrieved from http://firstmonday.org/ojs/index.php/fm/article/view/3336/2767.

Collins, M. (2014, May 9). #BringBackOurGirls: The power of a social media campaign. *The Guardian*. Retrieved from http://www.theguardian.com/voluntary-sector-network/2014/may/09/bringbackourgirls-power-of-social-media.

El País. (2011, December 26). *La ONU condena la violencia en Nigeria tras una cadena de ataques sectarios*. Retrieved from http://internacional.elpais.com/internacional/2011/12/26/actualidad/1324888448_092397.html.

El País. (2014, May 8). *Malala se suma a la campaña para la liberación de las jóvenes nigerianas*. Retrieved from http://sociedad.elpais.com/sociedad/2014/05/08/actualidad/1399531273_810291.html.

Fernández, C. (2014, May 10). Michelle Obama se muestra "indignada" por el secuestro de las niñas nigerianas. *El País*. Retrieved from http://internacional.elpais.com/internacional/2014/05/10/actualidad/1399731624_204741.html.

García, M., & del Hoyo, M. (2013). Redes sociales, un medio para la movilización juvenil. *Zer, 18*(34), 111–125.

Glenn, C. (2015). Activism or "slacktivism?": Digital media and organizing for social change. *Communication Teacher, 29*(2), 81–85.

Henríquez, M. (2011). Clic Activismo: redes virtuales, movimientos sociales y participación política. *Faro, 13*, 29–41.

Lyons, K., Robinson, W., & Chorley, M. (2014, May 8). *Bring Back Our Girls: Michelle Obama and Malala join campaign to free 276 Nigerian teenagers kidnapped by Islamic extremists*. Retrieved from http://www.dailymail.co.uk/news/article-2622999/Celebrities-join-campaign-bring-kidnapped-Nigerian-girls.html.

Morozov, E. (2009, May 19). The brave new world of slacktivism. *Foreign Policy*. Retrieved from http://foreignpolicy.com/2009/05/19/the-brave-new-world-of-slacktivism/.

Morozov, E. (2014). *To save everything click here: The folly of technological solutionism*. New York: Public Affairs.

Naranjo, J. (2014, May 5). Boko Haram amenaza con vender como esclavas a 223 jóvenes nigerianas. *El País*. Retrieved from http://internacional.elpais.com/internacional/2014/05/05/actualidad/1399304988_810578.html.

Neubauer, M. (2014, December 15). *TechPresident. Obtenido de #BringBackOurGirls: How a Hashtag Took Hold*. http://techpresident.com/news/24996/bringbackourgirls-how-hashtag-took-hold.

Resina de la fuente, J. (2010). Ciberpolítica, redes sociales y nuevas movilizaciones en España: El impacto digital en los procesos de deliberación y participación ciudadana, Mediaciones Sociales. *Revista de Ciencias Sociales y de la Comunicación, 7*, 143–164.

Vegh, S. (2003). Classifying forms of online activism: The case of cyberprotests against the World Bank. In M. Ayers & M. Mccaughey, *Cyberactivism online, activism in theory and practice* (pp. 71–96). London, UK: Routledge.

CHAPTER 9

How Does Politics Work? The Big Data View

The agents that form part of the digital public sphere operate in the ambivalence generated by disintermediation. As we have seen in previous chapters, there are examples that show how the disintermediation of the agents, spaces and messages have created a scenario in the which citizens show that they are able to react to events they consider acts of injustice or to claim their rights. We also find that these citizens are capable of creating their own communicative content that, thanks to the characteristics of Web 2.0, can become viral and make the whole planet a potential arena for public opinion.

Accepting its positive aspects, disintermediation also offers a dark side. We have pointed out in the previous sections how the disintermediation of agents can lead to the emergence of organizations that, although at first, they may seem more open and horizontal, use the participatory aspirations of citizens as an opportunity to obtain support or add votes. For its part, the disintermediation of messages brings with it an individualization of support for causes, claims and actions that, instead of being understood as a collective action, are taken as a personal affective affinity. Finally, the disintermediation of spaces has consequences of great relevance and should not be underestimated. Among them, the submission to processes governed by speed and, therefore, the lack of maturity and reflection, as well as the false idea that all problems have a solution and that this is unique and technical.

© The Author(s) 2019   115
J. M. Robles-Morales and A. M. Córdoba-Hernández,
Digital Political Participation, Social Networks and Big Data,
https://doi.org/10.1007/978-3-030-27757-4_9

However, other problems that, we believe, are just as important, hover over the potentials of disintermediation. In this chapter we wish to emphasize the elements that make this public opinion an arena with enough risks and challenges to make us prudent when diagnosing its benefits and potential. The ambivalence of this disintermediated public opinion is the consequence, as we have shown in the theoretical section of this book, of behaviour such as homogeneity, polarization, incivility and flaming. These attitudes are potential enemies of public opinion according to the positive terms that Habermas used to define it. Public opinion that as we pointed out should be potentially ready to be critical, reactive and independent.

As we shall see in the following sections, the use of Big Data techniques will allow us to advance in our understanding of the damaging effects of the digital development of public opinion and especially of disintermediation. Our case studies here will be the presidential elections in the United States and in Spain, both of which were held in 2016. The selection of these cases responds to our interest in studying processes linked to both conventional and unconventional political participation. Thus, if in Chapters 7 and 8 we analyse two cases of unconventional political participation, this time we focus on two cases of political partition in an electoral context.[1] In both cases, we shall study data from the Twitter social network by applying the Big Data techniques of sentiment analysis and text mining, by analysing the frequency of social behaviour such as polarization, incivility and flaming in public debates on such important political events in their respective countries.

9.1   THE NEGATIVE EFFECTS OF DISINTERMEDIATION: #TRUMP2016

One of the most salient features of Western liberal democracies like that of the United States is its notable tendency towards two-party systems. In other words, a political scenario in which the media and public opinion focus their attention on two political parties, therefore choosing between the two candidates that these parties put forward. In a context such as this, public opinion tends to be structured around two topical poles or central

---

[1] However, and as we pointed out above, the objective is not to extract lessons on the extent to which these two types of participatory contexts affect our research. For this, a different research design would be needed. Our only thing is to have different cases in which disintermediation is a central element.

themes as a result of the narratives of the main competing parties, as well as the media and pressure groups that support them. Therefore, it is to be expected that public debate inherits this same two-pronged structure and there is a certain level of polarization created between the supporters of one party and the supporters of the other. Sartori (2005) has referred to this type of scenario to remark on the importance of not confusing *polarization with polarization around parties*. While the second term refers, as we said, to the distance between supporters of one or the other party, the first refers to a process in which positions become more and more estranged.

Sartori says that a certain degree of alternance is to be expected in a two-party context, as parties vie for voters in the political centre or a centripetal trend. This last variable is the one the author considers most significant. The more centripetal the political tendency is, in the sense that the parties aim their discourses at the same point (the political centre), the more shorter the ideological distance between the parties is likely to be (polarization around parties). In contrast, if the tendency is centrifugal, so that the discourses of the parties generate more distance between them as they adopt irreconcilable ideologies or positions, polarization then becomes a problem for all political processes (negative polarization).

For many authors, the recovery from the financial crisis is being achieved at the expense of a reassessment of populist policies (Inglehart & Norris, 2016). Although this concept is complex and hard to define, we shall consider it here as a type of political discourse based on ideological simplification (good and bad, them and us, etc.), an anti-elite stance (speeches in favour of the disadvantaged, the dissatisfied, etc.), the role of a charismatic figure and/or a call to social mobilization (Laclau, 2005). This is clearly a form of discourse that creates centripetal tendencies while underlining differences with the opponent, basing the message on minorities or irrational issues such as charisma or the so-called "natural" quality of the candidate (Laclau, 2005).

Various authors have pointed out the important role that populism, as defined above, has played in the last presidential campaign in the United States (Kellner, 2016). Although there were several figures who stood out for their populist discourse, such as Democratic Party candidate for the presidency of the United States, Bernie Sanders, the main focus of attention was undoubtedly Donald Trump. There have been many studies that show the populist nature of the general discourse of the Republican candidate (Kellner, 2016). The analysis of Trump's speeches by Oliver and Rahn (2016), showed his repeated tendency to make use of a simplified message, establishing between the good and the bad, or an anti-elitist stance in which the crisis in the political system was the focus of his attention. Other authors have also noted the role of Trump as a charismatic figure who uses natural, everyday language to speak to the electorate as an equal.

Ultimately, we agree with Sartori (2005) when he says that polarization, when it occurs around parties, does not represent a political problem. Our uncertainty increases, however, when certain factors such as the growing importance of populist discourse mean that the tendency shifts in two-party systems away from a centripetal to a centrifugal one.

The other important factor in our understanding of the effects of polarization is its affective dimension. Literature has been telling us for decades about the effect of affective dispositions on membership, support and voting for a particular candidate. Similarly, affective polarization, as we saw in Chapter 2, has also been explained as a leading factor by experts in political communication (Lelkes, 2016). In this sense, polarization is visible in the (affectively directed) positions citizens take depending on their ideological positions and the discourse of the political candidates. We find once again, in this sense, that polarization is an integral part of a two-party system.

Many authors, however, have warned that it is a process that in many ways acts against this dynamic. As we saw in the theoretical framework for this book, the concept of "negative partisanship" has been coined to refer to the citizens' tendency to see themselves as politically hostile to a party or candidate instead of the traditional idea of supporting the one which best represents their opinions and interests (Abramowitz & Webster, 2016). This is how a significant number of United States citizens came to decide their vote on the basis of taking into account the type of result they didn't want (the policies of the candidates they reject), and the type of leaders that they didn't respect, instead of on the basis of the policies that they would like to see implemented.

"Negative partisanship" is the result, according to Abramowitz and Webster (2018), of the polarization of the discourse around racial issues in the United States, the proliferation of media clearly biased towards one or the other candidates, and also the value of "personality" in electoral processes.

Although this type of partisanship appears to share certain affinities with populist discourse, such as simplification (good and bad, whites, Hispanics and blacks, etc.), or the charisma of a personality, we are interested in treating it as a separate phenomenon that defines citizens' positions in negative terms. In other words, adopting a hostile attitude (hate) towards the opponent. This affective attitude is not, in principle, one which favours constructive public debate.

Studies have shown that both polarization around parties and affective polarization can be intrinsic components of the two-party system landscape. The presence of other phenomena such as populism or negative partisanship can alter this tendency within a two-party system so that this polarization becomes a factor which obstructs political communication.

In this section we shall try to understand the tendency of political debate on Twitter during the US presidential campaign on the premise held by various authors that this campaign was influenced by both populist discourse and by negative partisanship (Abramowitz & Webster, 2018; Inglehart & Norris, 2016). We are interested in examining the extent to which this scenario generated or did not generate (in general and affective terms) negative polarization, which would have included behaviour such as incivility and flaming.

In this sense, Table 9.1 shows the possible combinations of scenarios that we seek to identify, and each of them has different implications for political communication. Should we find, for example, moderate polarization in which citizens and other political agents offer arguments to support their preferred candidates, we would be in normal two-party partisanship

Table 9.1 Theoretical distribution of different types of polarization

		Nature of the polarization	
Political context of the polarization		Positive affective position	Negative affective position
	Polarization around parties	Expected in two-party systems (scenario 1)	Potentially harmful for the development in a communicative context (scenario 2)
	Polarization	Although to be expected, it indicates significant distance between positions among the electorate (scenario 3)	Clearly damaging because it shows a hostile electorate divided by its points of view (scenario 4)

Source Own research

situation that the system can cope with (scenario 1). In other words, what Sartori (2005) calls *Party Polarization*. Situations in which these political agents adopt affective positions in opposition to the candidates of other parties (scenario 2) or adopt positions that are so distant as to make compromise and debate difficult (scenario 3) are less desirable. We can base this assumption on the fact that political debate is more difficult when the starting point is the rejection of the opponent or two radically different positions. Finally, the most negative scenario of all is that in which we find both severe polarization and a very negative attitude towards the opposing candidate (scenario 4). This would be an unmistakable indication of a scenario where political communication has broken down.

For this study we downloaded, through the Twitter API, all the messages that contained direct allusions to the two candidates (@HillaryClinton and @realDonaldTrump). Dating from 0:00 (UTC) of 1 October 2016 to 9:00 (UTC) of 13 November 2016. The end of the election campaign and two days after the elections. We downloaded a total of 13,358,353 original messages and retweets, taken from 2,967,701 users.

To measure the polarization of public opinion we used the polarization index defined by Morales, Borondo, Losada, and Benito 2015. To study the feelings expressed we used the *supervised polarity classification* technique (Pang & Lee, 2008). We also used the *naïve Bayes classifier* (Rish, 2001) implementing and using the library *Spark MLlib*. With all these we classified the messages as follows: Positive (love) for D. Trump, Negative (hate) for D. Trump, Positive (love) for H. Clinton, Negative (hate) for H. Clinton or Neutral.

We were able to use these techniques to create a love-hate diagram that showed the position of the different agents taking part in the public debate surrounding the two candidates. Each section of this diagram has a different meaning in terms of attitude towards each of the candidates (Clinton-Trump) as described in Table 9.2.

Table 9.2 Love/hate for each of the candidates

Quadrant	C	T	Meaning (Clinton-Trump)
Q1	>0	>0	Love–Love
Q2	>0	>0	Hate–Love
Q3	>0	>0	Hate–Hate
Q4	>0	>0	Love–Hate

Source Own research

The coordinated axes $C = 0$ and $T = 0$ imply indifference towards Clinton and Trump respectively and the origin $(0,0)$ indifference towards both candidates. The individuals who are on the $T = C$ diagonal of the diagram have the same attitude towards both candidates, while the $T0$-C diagonal implies that there is as much love/hate for one candidate as there is love/hate for the other, representing the users with most "polarized" opinions. The nearer we find users to the ends of this line, the more polarized their opinion of one candidate in relation to the other. Point $(-1, +1)$ implies complete Hate for Clinton and complete Love for Trump. In contrast, point $(+1, -1)$ implies complete Love for Clinton and complete Hate for Trump.

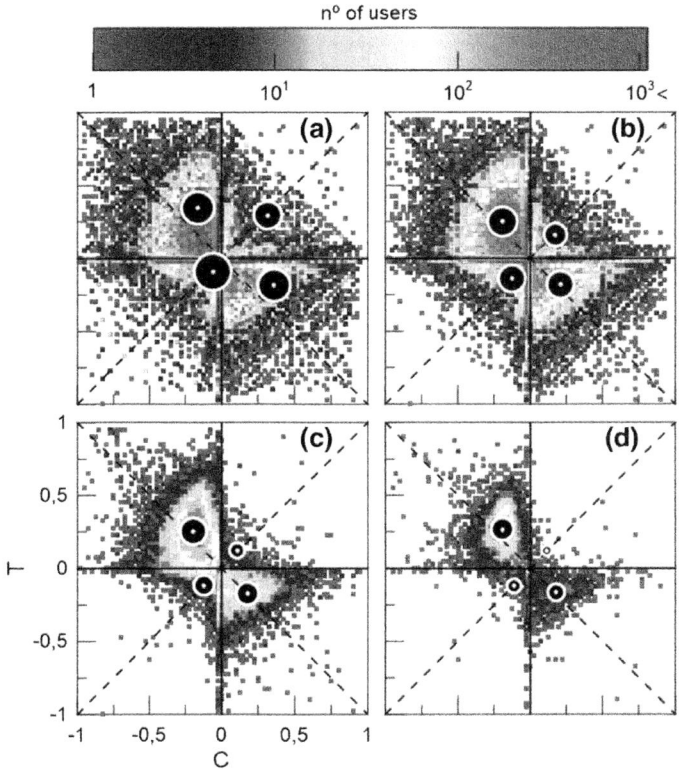

Graph 9.1 LOVE-HATE diagram by user activity (*Source* Own research)

Graph 9.1 shows the LOVE-HATE diagram in relation to user activity. The diagram shows the average opinion of the users who are found in each of the quadrants. Information is shown about:

(a) all users
(b) users who have written 10 tweets
(c) users who have written 50 tweets
(d) users who have written more than 150 tweets.

The colour indicates the number of users according to the scale shown at the top of the graph. The black dots show the average opinions of each quadrant and their size is proportional to the logarithm of the number of users in the quadrant.

These diagrams enable us to arrange the information so that the results can be interpreted intuitively. For a clearer explanation, we need to remember that the lower and further to the left the messages are (dots in the diagram), the more uncivil and inflammatory the messages for both candidates are. The messages that are at the top-right are positive for both candidates. The messages in the top left quadrant are positive for Trump, while those in the lower-left quadrant are positive for Clinton.

Graph 9.2 shows the LOVE-HATE diagram expressed in absolute terms:

(a) For all users who wrote fewer than 10 tweets: 1,260,477 in total.
(b) Users who have written more than 10 tweets: 147,336 in total.
(c) Users who have written more than 50 tweets: 34,515 in total.
(d) Users who have written more than 150 tweets: 8110 in total.

The colour indicates the percentage of users according to the scale shown at the top of the graph. The black dots show the average opinions of each quadrant and their size is proportional to the logarithm of the number of users in the quadrant.

As we have said, political polarization is a topic of vital importance in understanding current political processes. This situation has generated even more interest as a consequence of the political development of the Internet and its potential impact on the polarization of attitudes, relations and sources of information. This work has focused on the study of political polarization of debate in Twitter in one particular electoral process; the last presidential campaign for the United States in 2016.

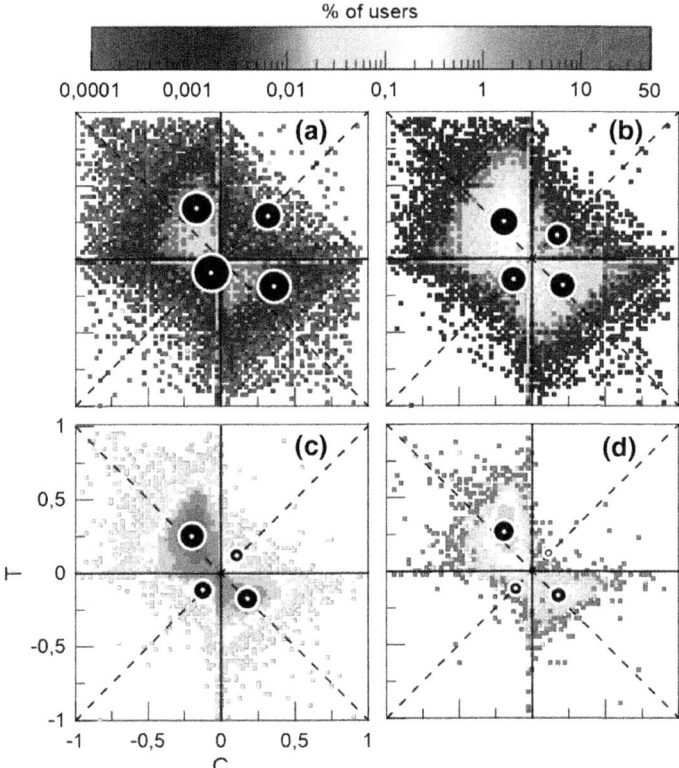

Graph 9.2 LOVE-HATE diagram for all users (*Source* Own research)

We have adopted the premise that a certain amount of polarization is to be expected in two-party electoral systems: polarization around parties (Sartori, 2005). However, various authors have affirmed that this tendency can be altered by the recurrent of phenomena, like populist discourse or negative affectivity, that generate scenarios in which public debate diverges from an ideological centre and becomes more difficult (Abramowitz & Webster, 2018; Inglehart & Norris, 2016).

In methodological terms, we have used data from Twitter that was downloaded during the electoral campaign in the United States using the two main candidates, Hillary Clinton and Donald Trump as reference points. We have used this data to calculate the degree of polarization and the affective disposition of the actors in relation to each of the candidates, in order to test whether the level of polarization to be expected from a two-party scenario (polarization around parties) did actually arise in the debate on Twitter during an election campaign under the circumstances indicated above.

Our results lead us to reach several conclusions. First of all, the degree of polarization measured using the technique of Morales et al. (2015) was not high (0.2436). We also observed that most participants in the debate maintained a neutral position, meaning that they were equally disposed to Trump or Clinton. In contrast, extreme positions barely existed, while the most likely value in each interval had the absolute value of 0.5 ($-0.5 < X < +0.5$). A truly moderate value. Given these results, we can say that although the context of the elections was marked by factors that could potentially have altered the situation of polarization around parties (Sartori, 2005), there are indications that this did not happen in the Twitter debate on the two rival candidates.

Secondly, the participants do not show extreme affective dispositions. We can see in Graphs 9.1 and 9.2 how the space in the bottom left quadrant that is reserved for users who display negative affective disposition to both candidates is very close to zero. This means that there is no "negativized" affective disposition. In contrast, we can see in the upper right quadrant, which shows users who express positive values about the candidates, that the centre of gravity is much closer to 1 (messages with positive tone). When it comes to the affective position in relation to each candidate (diagonal from left to right), we can see a tendency that we expect to see in polarization around parties. In other words, a moderate affective polarization in which the defenders of each candidate base their opposing views on their electoral preferences. This result argues against the thesis that electoral campaigns generate negative affective polarization in debates that take place on Twitter.

In short, our data seems to fit best with scenario 1 from Table 9.1, where we showed the possible combinations of the results from our study, with the differences in each scenario in terms of political effects. The low polarization (0.2436) and relatively positive attitude towards the candidates place our results in the least negative scenario, which coincides with what we would expect in two-party systems. We can speculate here about the effect that the type of context has on the communication strategies of the agents. We consider that despite the potentially "polarizing" ingredients present in the context, the agents, especially citizens with no links to organizations, feel that they are in a formal situation (a political debate) in which moderation should prevail. We know that the radicalization of messages is more plausible in less formal and more spontaneous situations.

However, this description fits perfectly with the case in which all the users are included in the graph. It is important, however, to stress that removing the users who were less active in the debate from the graph reveals a degree of radicalization. In figures "b", "c" and "d", in Graphs 9.1 as well as 9.2, the centres of gravity show a significant movement downward and to the left. This displacement suggests an increase in messages with negative content and therefore an increase in the negative affective disposition of those responsible for these accounts on Twitter.

If we accept the premise, that the more active accounts in this type of process belong to organizations such as parties and/or the media, we can consider the idea that tendencies vary in accordance with the type of Twitter account. This means that the less active users, who are presumably citizens with no links to organization, show an affective disposition that is less polarized than that of more active users. This means that polarization would increase in proportion to the activity of the Twitter account.

Thanks to these results we can say that, in general, polarization is relatively moderate despite a potentially negative context. We can qualify this by affirming that the debate acquires a more negative tone when we focus on the most active accounts. This result may be a consequence of intentional actions by these agents to intensify the debate, although this need not be the case, by including inflammatory press releases or highly negative declarations on the part of politically relevant figures. Even so, this question lies beyond the objectives of this book and would be the subject of future developments in our research.

9.2   Nodes that Win Elections: #UnidosPodemos

In this section we have chosen[2] the case study of the political party Unidos Podemos. Specifically, we have selected the political debate that took place in Spain on digital social media (Twitter) around the candidacy of this party in the General Elections for the Government of Spain in June 2016. We chose this party based on the following considerations. First of all, the party emerged from the political debate that arose in Spain around the

---

[2]A more complete version of this study has been published in Information, Communication and Society, 2018 (https://doi.org/10.1080/1369118X.2018.1499792).

15-M movement which, like other processes such as *Occupy Wall Street* or *The Arab Spring*, were central to global collective action in the first years of this decade (Romanos, 2016). Secondly, this new party generated a major political debate in Spain and new hopes for the political left in the country (Ganuza & Ganuza, 2014). Finally, the communication strategy of this party combined the deliberate reliance on digital technology with continuous presence in the mass media (Fernández, 2014). It is an example of a connective party, and therefore a party for which the disintermediation of agents is an important factor. All this makes Unidos Podemos an ideal case study for the subject of this book.

The objective of this section is to provide empirical evidence of the affective nature of political debate and its polarizing and uncivil effects. Specifically, we are asking to what extent the involvement of agents[3] in political debate is related or not with more or less fragmented political effects and the extent to which the debate topics are related with different types of affectivity. Taking the concept of "Affective Polarization" (Bafumi & Shapiro, 2009; Hetherington, 2001; Lelkes, 2016) as a reference point, we looked to see if there were emotive differences between agents with high, medium and low activity in online political debate, and at a total of five topics for each of these types of agents.

Our hypothesis is that when agents debate with other agents with different sensibilities than their own (positive + negative + neutral) we are in an affective scenario with relatively little affective polarization. If, to the contrary, we find debates in which participants only share opinions with agents of the same affective type (negative + negative; positive + positive), we are in an affectively polarized scenario. Finally, we cannot identify "affective polarization" if the agents engage in political debate with a neutral affective disposition because we must assume that there is no polarized attitude in these cases. In this case we shall introduce the dimension of topics as a control variable, to test the extent to which the issue in question influences the affective disposition of the agents.

Taking the concept of "Affective homogeneity" as a reference point, we looked to see if there were emotive differences between agents with high, medium and low activity in online political debate, and at a total of five topics for each of these types of agents. The degree of online political activity

---

[3] As explained in the methodological section of this article, our unit of analysis were the issuers of messages. Given the diversity of issuers in a context of political debate on Twitter, we chose to use the generic term "agent".

is a shared attribute that reveals the power or ability to spread information and to be present in the digital public sphere. We can classify the position of each agent in relation to a group of debate topics depending on whether their affective position is positive, negative or neutral, and analyse the extent to which there is homogeneity. Our hypothesis is that strong homogeneity connected with shared attributes is a gateway to political polarization.

The findings of this analysis are not only relevant to the study of the relationship between politics and the Internet, but go further and force us to ask vital questions about the democratic quality of developed democracies and the extent to which today's societies are more or less democratic (Sunstein, 2017). Homogeneity in political topics and, in wider terms, polarization is, in this light, a considerable hindrance to the proper development of political debate in the Network Society.

Some Methodological Key Ideas to Understand Polarization

This point will offer a brief description of the methodology applied in this study. First of all, it explains classification of the topics in terms of the authors that formed part of the social network and who issued a tweet that contained the two words UNIDOS and PODEMOS during the electoral campaign between 10 and 27 June 2016. The nodes of the network correspond to different users who were active during the electoral campaign. An arrow between two twitter users i and j so that it points from j to i means that j is mentioned by i, or that i has retweeted j. In this case, it means that the information has passed from j to i, so the direction should be shown as from j to i. If this situation is repeated during the electoral campaign, the value of the relation represents the number of time j is retweeted or mentioned by i, and therefore the increase in the influence of j over i.

It is important to note that the classification suggested does not occur at the level of the tweet but at that of the author, because we seek to understand how users of the network participate in relation to the topics that the messages deal with, and their affective dispositions. This approach is referred to in the literature as the author topic model. Hereafter we shall include all tweets issued by a user during the period of analysis under the term "document" and talk equally about classifying authors and documents.

As a starting point we have considered it appropriate to classify topics depending on the amount of activity by the agents. We will consider that the identification of a user with a topic is questionable when it is based on a single tweet. These cases have therefore been eliminated when identifying topics. We also decided to treat users who wrote masses of tweets differently, with the idea that these users probably covered all the topics. This means that we have segmented the corpus into agents with low, medium and high activity.

After this prior grouping we have carried out a data mining analysis on the group of texts (corpus) associated with each of the groups of users. This analysis consists of the following phases:

– Natural Language Processing (NLP)—which is intended to structure the data so it can be processed using an analytical model, reducing the size of the problem by eliminating irrelevant information and making transformations that seek to improve the results of the classification of contents.
– Identification of topics and classification of users based on these.
– Sentiment analysis—the purpose of which is to label users as positive, negative or neutral in regard to the polarity of the contents they publish.

The *Contextual Analysis* tool by the SAS Institute was used for each of these stages, because it is software that has been specifically designed to solve problems of this type.

The Types of Political Agents: What We Know About Political Polarization in Unidos Podemos Political Debate?

This section aims to show the results of the analysis carried out in this article. To do so, we shall structure the information as follows. First of all, we offer information about the affective disposition of each type of agent in relation to each of the topics that were most actively debated during the electoral campaign. Secondly, we shall summarize the topics covered by each type of agent (*high-active agents, middle-active agents* and *low-active agents*) and show a series of "representative" messages for these topics (Graph 9.3).[4]

Graph 9.4 shows the percentages of positive, negative and neutral agents in relation to each of the subjects discussed during the electoral campaign

---

[4] Unfortunately, due to the length of this work, we cannot summarize all of the subjects covered. The selection of topics to be summarized here is intended to be representative of the discourse of each type of agent.

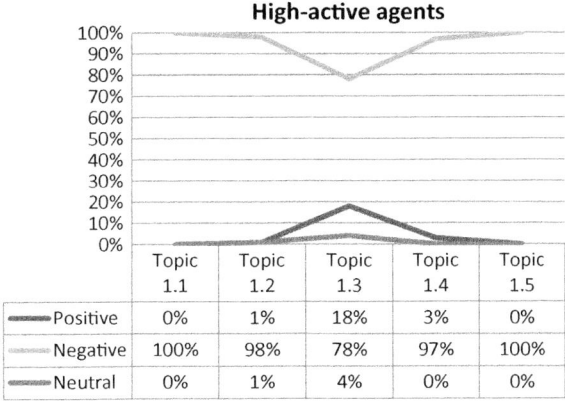

Graph 9.3 Affective position of the high-active agents depending on the subject of debate (*Source* Own research)

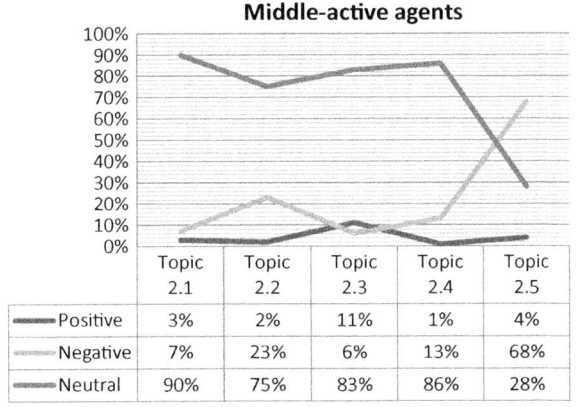

Graph 9.4 Affective position of the middle-active agents depending on the subject of debate (*Source* Own research)

around the candidacy of Unidos Podemos for the presidency of the Spanish Government.

We shall see that there is a strong majority of negative positions. In fact, 100% of the agents who took part in the debates on points 1 and 5 show this type of affective disposition. 98 and 97% of the participants in debates on topics 2 and 4 showed, respectively, this same disposition. The percentage of agents with a positive disposition is only significant for topic 3 (18%), although the negative disposition is clearly dominant here too.

Agents with a middle intensity use of social media when debating the candidature of Unidos Podemos for the presidency of the Spanish Government have a less distinct position than the most active agents (Graph 9.3). In this case, as we see in Graph 9.5, the most common disposition, except for the subject covered as Topic 5, is neutrality. The clearest case in which neutrality in the common ground is topic 1, in which 90% of the agents adopt this position. In the other subject, apart from topic 5, the percentage of agents with a moderate discourse was 75% (Topic 2), 83% (Topic 3) and 86% (Topic 4).

The debate engaged in by agents with less communicative activity also displays peculiar characteristics. In this case, Graph 9.5, we find that apart from the subjects covered in Topic 1, it is normal for the percentage of agents with negative and neutral dispositions to be very similar. At the same time, the percentage of agents with a positive disposition is above 10% in all cases. Finally, the majority position for Topic 1 is a negative affective disposition. Even so, roughly a quarter of the agents involved in this debate have a disposition that varies from the majority. In this case, they have a neutral or positive disposition.

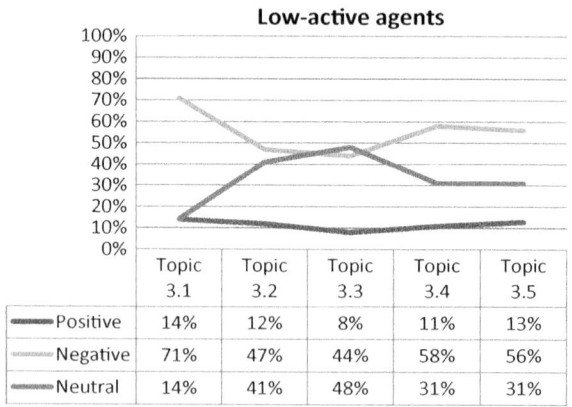

	Topic 3.1	Topic 3.2	Topic 3.3	Topic 3.4	Topic 3.5
Positive	14%	12%	8%	11%	13%
Negative	71%	47%	44%	58%	56%
Neutral	14%	41%	48%	31%	31%

Graph 9.5 Affective position of the low-active agents depending on the subject of debate (*Source* Own research)

To show the content of the debates of the high-active agents, we would like to highlight Topics 3 and 5. Topic 3 is important because it is the only one where there is a relatively high percentage of agents who have an affective disposition that is distinct from the majority (negative).

The central issue of Topic 3 is the start of the electoral campaign. It therefore deals with debates about the first public speeches and events. This topic, which is potentially neutral and free of radicalization (the campaign is just starting and debate has barely begun) has a negative tone when the agents focus the start of the campaign on criticism of the policies of the Partido Popular (PP), who were then in power.

The representative messages for this Topic (suggested by the software used for analysis) that we shall show below contain constant negative allusions and criticism of the PP right at the start, as we noted, of the electoral campaign.

Message 1: The priority for Unidos Podemos is to put an end to the damaging policies of the PP.

Message 2: Every time we hear the PP say that they defend motherhood, we are stunned.

Message 3: Despite the financial cutbacks by the PP, not only has expenditure not fallen, it has actually risen.

Topic 2 is also interesting because, like topics 1, 4 and 5, it brings around 100% of the agents with a negative disposition together on the subject under debate. The topic here is critical of the leaders of Unidos Podermos for suggesting that those attending party meetings do not carry communist flags. The Twitter users reject this strategy and the social-democratic discourse that Unidos Podemos has used throughout the electoral campaign.

Samples of this attitude, as recommended by the software, are:

Message 1: IU recommends not taking communist or republican flags to Unidos Podemos meetings.

Message 2: Electoral strategy of Unidos Podemos. Let's hide who we are so they won't know who they are voting for.

Message 3: The funny thing is that the "podemites" think we are crazy for saying that Unidos Podemos is communist.

It is important to note that although the general disposition in these two debates is negative and critical, the target of this negative affective disposition is, in these cases, different. While in case 1 the target of criticism is the Partido Popular (PP), in Topic 3 the target of the reproach is Unidos

Podemos. This leads us to think that the ideological profile of the agents in these debates are different, closer to Unidos Podemos in the first case and closer to the centre-right parties in the second. However, in each case, the agents are debating with people of a similar affective disposition.

To describe the topics that this type of agent discusses, we have selected topics 1 and 5 because they represent the content in which there is the greatest and least distance between the main affective dispositions: negative and neutral. Topic 1 basically gathers messages where Twitter users make predictions about the electoral results on election night. In this way, and taking the results from the opinion polls conducted at the polling stations and/or the official results offered by the government, the agents must debate the possibility of Unidos Podemos finishing in second place in the general elections for the presidency of the Spanish Government or whether, to the contrary, they will remain in third place behind the Partido Socialista Obrero Español (PSOE).

Some examples of the messages sent by the Twitter users and suggested by the software for this analysis are:

> *Message 1:*　with 82% of the votes tallied, the PP has 136 seats, PSOE 88, Podemos 71 and Ciudadanos 30.
>
> *Message 2:*　the CIS says PP 29 [seats], Unidos Podemos and associates 25, PSOE 21, Ciudadanos 14.
>
> *Message 3:*　Opinion poll by Sigma Dos PP 117-121 [seats], Unidos Podemos 91-95, PSOE 81-85 and Ciudadanos 26-30.

In contrast, topic 5 shows a more negative tendency as a result of seeing that the Partido Popular (PP) has won the elections with more votes than in the previous election. In this case, the negative disposition of the debate increases until it reaches practically seven out of every ten agents. This disposition is mostly due to the fact that the different opinion polls during the whole election campaign had forecast that Unidos Podemos would finish in second place and especially that the Partido Popular would lose seats. These results had been interpreted as the electorate punishing the government for its handling of the financial crisis and the cases of political corruption. When this did not occur, the agents took advantage of Unidos Podemos' poor results to attack the party.

Some examples of this position are:

Message 1:   PP increases its votes and Podemos loses a million and a half
                 votes. Chavismo[5] is losing on all sides
Message 2:   el sorpasso[6] has shown the strength of Unidos Podemos los-
                 ing more than a million votes in comparison with 20d
Message 3:   unidos podemos have clearly failed by keeping the same num-
                 ber of seats while losing over a million votes

We can see how the description of the electoral results in the first topic
was seen from a neutral perspective, while topic 5 is clearly a criticism of
Unidos Podemos based on its electoral results. Despite this, the agents
with a negative disposition share a lower, although substantial, percentage
of this debate with those of a neutral affective disposition. This leads to a
degree of interaction between these affective dispositions.

To describe the debates between the low-active agents we shall take
topics 1 and 5 as examples. While topic 1 shows the clearest difference
between the two leading affective dispositions (negative and neutral), topic
5 is an example of a debate in which the three affective dispositions are
present in significant percentages.

The subject of topic 1 refers to the ideological struggle between the
main Spanish media outlets in relation to the political options of Unidos
Podemos. Participants also mention the main supporters of the government
to show that Unidos Podemos faces an adversary in a very strong position.
In this scenario, the agents who took part in the debate took advantage
of the context of the discussion to reinforce the contents of the Unidos
Podemos manifesto.

The software proposed the following messages to sum up the debate:

Message 1:   in the last days of the campaign and the attacks on Unidos
                 Podemos from those who have pillaged and ravaged Spain
                 are intensifying
Message 2:   My God! Merkel Obama PP the church PSOE CDC and
                 Ciudadanos with all the financial and media power all against
                 UnidosPodemos
Message 3:   bearing in mind the editorial line taken by all the press, the
                 success of UnidosPodemos will be remarkable, if confirmed

---

[5]One of the main attacks on Unidos Podemos during the electoral campaign was the
proximity of some of the party leaders to the *chavista* regime in Venezuela.

[6]This expression was used during the campaign to describe the rise of Unidos Podemos in
the elections and its victory over the second political force in Spain, the PSOE.

Topic 5 revolves around the position of one of the main leaders of Unidos Podemos, Iñigo Errejón, who was then the party's number two. In the view of this politician, the merger of Podemos with Izquierda Unida (Unidos Podemos) could have the opposite effects to the intended ones, losing dissatisfied supporters from both parties instead of obtaining more votes. Given the electoral results, this may have happened, and the agents taking part in debate 5 saw it this way. This debate also includes other general topics related to Unidos Podemos as a solution to the political problems of Spain.

Some examples of this position are:

> *Message 1:* He did it again [Errejón] has hit the nail on the head on why
>   UnidosPodemos did not win the elections
> *Message 2:* Errejón was right
> *Message 3:* people have lost so many things that they also lost their fear
>   UnidosPodemos

We can find, in the debate between this last type of agent, positions that are favourable to Unidos Podemos. Even so, the affective disposition of the agents taking part is plural. The most common affective stances are negative and neutral ones, however. Still, there is an important percentage of agents with a positive disposition. We can therefore find a scenario in which the agents have to debate with others with distinct affective dispositions.

In this chapter we are particularly interested in seeing whether affective polarization, if present, is expressed more clearly among the political agents who are more or less committed to public debate on digital social media. We would also like to know if the topic of debate was in some way related to this affective polarization.

Once finished the empirical analysis we can extract meaningful conclusions for our purposes. First of all, we know that the negative affective disposition is clearly predominant in the debates in which high-active agents took part. The debates in which middle-active agents took part had a predominantly neutral affective disposition. Finally, the low-active agents took part in debates in which actors with negative disposition shared predominance with those who had a neutral disposition.

As we established in the introduction to this section, we understand that a scenario a) defined by a mixture of affective dispositions or b) dominated by agents with a neutral disposition cannot be described as "polarized". We consider, along with Abramowitz (2010), Fiorina and Samuel (2008) or Abramowitz and Saunders (2005) that polarization can be defined as the distancing of agents' points of view or their reinforcement in incompatible positions. In contrast, if an agent is exposed to opinions and content issued by participants with a different affective disposition, we can assume that the circumstances outlined by these authors do not apply. In this case, we consider that the scenario in which middle and low-active agents participate is not marked by affective polarization. On the other hand, the debates in which high-active agents take part, distinguishable by the hegemony of a single affective disposition, do, in our view, meet the criteria established for definition as affectively polarized.

This situation is confirmed if we observe that the high-active agents show a homogeneous (negative) affective disposition in relation to the debate topic. We can therefore see that a negative affective position dominated both debates where the agents discuss Unidos Podemos (left-wing party) and debates in which the party criticized is the Partido Popular (main right-wing party and the main political adversary of Unidos Podemos). This result is consistent with those given by the literature. As we saw earlier (Boxell, Gentzkow, & Shapiro, 2017; Davis & Dunaway, 2016), we know that citizens who are more informed and more interested in politics, and who therefore are more likely to be politically active, can adopt more polarized positions. In our case, adding a variable that is rarely used in the literature, we saw that commitment to political debate can have an important influence on affective polarization.

The fact that the most active agents on social media are the most polarized in affective terms has important implications. The clearest of these being that more active and committed debate seems to mean that debate in general, online or off-line, is polarized and, in this case, negative. The visibility achieved by the most active agents may lead us to think that we are looking at a fragmented political arena when this disposition need not be the most common among the majority of citizens, and not even among the agents taking part in the online debate who are less active. In the same vein, we know that the most active agents in online debates are usually institutional agents (political parties, media, etc.). If this is so, we are in a process of polarization whose main drivers are precisely those agents, like the political parties or the media, who play a central role in representative political systems.

A secondary, but not minor, aspect to consider is that despite the affective polarization that dominates the debate between the *high-active agents*, we do not find the kind of extreme radicalization expressed in clearly negative behaviour such as *incivility* (Rowe, 2015) or *flaming* (O'Sullivan & Flanagin, 2003). These are more common in the discourse of the most active agents and, in contrast, practically unknown among middle and low-active agents.

In short, we feel that the results obtained in this study provide evidence that can inform another, wide-ranging debate. We're referring to the debate about the effects of polarization on the proper development of the deliberative dimension of democracy. More empirical evidence is required to support the thesis of polarization in general, and of affective polarization in particular. However, studies such as this one provide us with evidence about a dangerous trend for the principles that ought to guide public debate, namely those of reason, impartiality, deliberation and publicity (Estlund, 1997).

9.3   SOME CONSIDERATIONS ABOUT THESE RESULTS OF THIS STUDY

The cases explained in this section are merely examples of a much broader process. The important thing in both cases is that with different statistical techniques and political contexts that are notably different, we can find similar patterns that support our outlining of some key ideas. Naturally, these results are open to question through the presentation of other different cases in which the tendencies identified may be different.

To date, we have found that the electoral processes we have studied, despite their differing contexts, show polarization and negative attitudes expressed in messages that do not contribute to the creation of public debate (in many cases these messages are impolite and are solely intended to inflame any debate). This is clear, for example, in the messages positioned in the lower left quadrant of Graphs 9.1 and 9.2 when, discussing both candidates, a significant portion of the electorate does so with critical and insulting messages.

In the same way, we find polarization present, but in the observed cases this could be attributed to polarization around parties (Sartori, 2005) but not to negative polarization. It intensifies as we restrict our view to the accounts that are most active in both electoral campaigns. While smaller and less active accounts tend to express positive or neutral opinions, those who are in charge of the most active accounts usually express more polarized and negative (potentially also less constructive and respectful) opinions.

Some authors have indicated that the most active accounts usually belong to organizations and the media. It is more common for citizens who are not members of these organizations to have a less active presence in public debate. In light of the results presented here, this fact can lead us to conclusions that are of interest for this book.

First of all, polarization and other behaviour that we have defined as negative is present in public online debate. This is naturally an obstacle to the construction of public opinion. However, this type of behaviour is most commonly found among the organizations that have traditionally occupied the public sphere. Smaller accounts (potentially in the hands of citizens) show a more neutral and/or positive polarized behaviour. Even so, the prominence of these traditional agents (media, political parties, etc.) makes the whole of public opinion appear to be polarized. The reality, however, is that it is precisely those agents who do not represent a scenario of disintermediation that indulge in behaviour that is less compatible with the correct formation of public opinion. Finally, if we were able to obscure the presence of large accounts in an environment that is closer to the ideal of disintermediation, public opinion would appear to be more open and positive.

This is, of course, merely a speculative exercise, and the reality shown by our research is that digital public opinion is hard to define because of the predominant presence of the traditional actors who polarize and raise the temperature of the debates. This creates, as we shall discuss at greater length in our conclusions, a scenario that is far from the ideal of public opinion that we discussed in this book when examining the ideas of J. Habermas.

References

Abramowitz, A. I. (2010). *The disappearing center: Engaged citizens, polarization, and American democracy*. New Haven, CT: Yale University Press.

Abramowitz, A. I., & Saunders, K. L. (2005). Why can't we all just get along? The reality of a polarized America. *The Forum: A Journal of Applied Research in Contemporary Politics, 3*, 1–22.

Abramowitz, A. I., & Webster, S. (2016). The rise of negative partisanship and the nationalization of US elections in the 21st century. *Electoral Studies, 41*, 12–22.

Abramowitz, A. I., & Webster, S. (2018). Negative partisanship: Why Americans dislike parties but behave like rabid partisans. *Political Psychology, 39*, 119–135.

Bafumi, J., & Shapiro, R. Y. (2009). A new partisan voter. *The Journal of Politics, 71*(1), 1–24.

Boxell, L., Gentzkow, M., & Shapiro, J. M. (2017). *Is the Internet causing political polarization? Evidence from demographics*. Cambridge, MA: National Bureau of Economic Research.

Davis, N. T., & Dunaway, J. (2016). Party polarization, media choice, and mass partisan-ideological sorting. *Public Opinion Quarterly, 80*(S1), 272–297.

Estlund, D. (1997). Beyond fairness of deliberation: The epistemic dimension of democratic authority. In J. Bohman & W. Rehg (Eds.), *Deliberative democracy: Essays on reason and politics* (pp. 173–204). Cambridge, MA: Massachusetts Institute of Technology.

Fernández, J. (2014). ¿Quién votó a Podemos? Un análisis desde la sociología electoral. In. F. M. González (Ed.), *#Podemos. Deconstruyendo A Pablo Iglesias* (pp. 75–101). Bilbao: Deusto.

Fiorina, M. P., & Samuel, J. A. (2008). Political polarization in the American public. *Annual Review of Political Science, 11,* 563–88.

Ganuza, I., & Ganuza, E. (2014). ¿Podemos? La Ametralladora. Madrid, Spain.

Hetherington, M. J. (2001). Resurgent mass partisanship: The role of elite polarization. *American Political Science Review, 95*(3), 619–631.

Inglehart, R., & Norris, P. (2016). Trump, Brexit, and the rise of populism: Economic have-nots and cultural backlash (Working Paper).

Kellner, D. (2016). *American nightmare: Donald Trump, media spectacle, and authoritarian populism.* Rotterdam: Sense Publishers.

Laclau, E. (2005). *La Razón Populista.* Buenos Aires: Fondo de Cultura Económica.

Lelkes, Y. (2016). Mass polarization: Manifestations and measurements. *Public Opinion Quarterly, 80*(S1), 392–410.

Morales, A. J., Borondo, J., Losada, J. C., & Benito, R. M. (2015). Measuring political polarization: Twitter shows the two sides of Venezuela. *Chaos, 25*(3), 33114, 1–9.

O'sullivan, P. B., & Flanagin, A. J. (2003). Reconceptualizing 'flaming' and other problematic messages. *New Media & Society, 5*(1), 69–94.

Oliver, J. E., & Rahn, W. M. (2016). Rise of the Trumpenvolk: Populism in the 2016 election. *The ANNALS of the American Academy of Political and Social Science, 667*(1), 189–206.

Pang, B., & Lee, L. (2008). Opinion mining and sentiment analysis. *Foundations and Trends in Information Retrieval, 2*(1–2), 1–135.

Rish, I. (2001). An empirical study of the naive Bayes classifier. In *IJCAI 2001 Workshop on Empirical Methods in Artificial Intelligence* (Vol. 3, No. 22, pp. 41–46). New York, NY: IBM.

Romanos, E. (2016). De Tahrir a Wall Street por la Puerta del Sol: la difusión transnacional de los movimientos sociales en perspectiva comparada. *Revista Española de Investigaciones Sociológicas, 154,* 112–122.

Rowe, I. (2015). Civility 2.0: A comparative analysis of incivility in online political discussion. *Information, Communication & Society, 18*(2), 121–138.

Sartori, G. (2005). *Parties and party systems: A framework for analysis.* Colchester, UK: ECPR Press.

Sunstein, C. (2017). *#RepublicDivided democracy in the age of social media.* Princeton, NJ: Princeton University Press.

CHAPTER 10

Conclusions

In this book our intention was to discuss the configuration of public opinion in a context of disintermediation of political communication. In this sense, we have proceeded moved by the so-called "Coleman's boat" (Coleman & Coleman, 1994). According to this image, for a correct explanation, the researcher must, first, refer to the level of the social structure (in our case, the communicative structure) to, thanks to it, establish the basic conditions. Then, we have to unravel the mechanisms by which these conditions are transferred to the individual behavior of the agents. Thirdly, it has to specify how, from the interaction between agents, new behaviours are produced that, in the fourth place, return to society variations with respect to its original structural organization. A graphic representation of the system will clarify our way of proceeding (Graph 10.1).

The starting point of our scheme is disintermediation. This process is taken as a theoretical assumption. That is, we do not submit it to any empirical evidence. On the contrary, we assume, summarizing the existing literature, that it is a recurrent phenomenon in the technologically developed communicative contexts. In this sense, we have explained that disintermediation is related to the effects of Web 2.0 technology on communication, in general, and on political communication, in particular. Thus, the reduction of costs of production of cultural content that generate this type of technology has an effect, at least so expressed by authors such as Benkler

© The Author(s) 2019
J. M. Robles-Morales and A. M. Córdoba-Hernández,
Digital Political Participation, Social Networks and Big Data,
https://doi.org/10.1007/978-3-030-27757-4_10

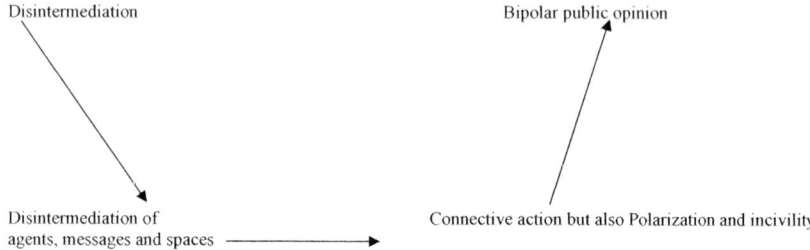

Graph 10.1   Coleman's boat for disintermediation process (*Source* Own research)

(2006) or Castells (2009), on the ability of citizens to establish a communication system in which traditional mediators, read the media or political parties, play a less decisive role compared to the communication systems of industrial society. This does not mean the disappearance of traditional agents, but their need to adapt and share space with other new mediators that break the hegemonic position of those. Being that way, mass self-communication (Castells, 2009) informs us about the ability of citizens to organize horizontally and, following Benkler (2006), to establish a real time, autonomous and global communication system.

However, mass self-communication and the disintermediation of political communication are "unforeseen consequences" of technological development. This concept coined by the sociologist R. Merton (1910–2003) refers to those processes that are not imagined by the original intention of the social actor. That is, the inventors of the Internet did not develop this technology with the intention that citizens would fight in the public space with media, political parties or social movements. It is the appropriation (specific actions) that citizens make of this technology, which generates this effect (disintermediation).

The appropriation of technology for the expression of demands and political opinions, as well as to coordinate in a connective action, is an innovative process that obeys, to a large extent, the interest of citizens to find a scope of expression and participation. The unforeseen consequence is that, the development and extension of digital technologies, becomes a structure of political opportunity for the citizens of democracies with great problems of public disaffection.

However, what form does that appropriation take? In other words, if we go down from the structural level in which the process of disintermediation is found, what things do citizens do? In this book we have pointed out the three that seem most relevant to us and we have analysed them with several recent examples. These three processes are the disintermediation of agents, messages and spaces (the bottom left of the boat).

We find how, the context of disintermediation favours the emergence of connective parties (births that establish, through a mixture of assembly and digitally mediated communication, a system of outsourcing of organizational functions previously reserved to the elite of the party), as well as a flexibilization of the relations between citizens and the administration and government systems of developed countries (e-government and e-administration). The connective parties, although also the e-administration, are examples, from our point of view, of how the organizations take the disintermediation to daily and concrete practices of the political communication. Open the internal debate of political parties to the direct participation of citizens is to break with asymmetric relations that were traditionally based on the idea of representation of the will of the party's supporters. A mediation, this one, that was sometimes so subtle that it seemed completely turned away from public opinion.

However, we observe that disintermediation permeates other areas of concrete and daily communication of citizens. Disintermediation, favored by tools such as digital social networks, implies that citizens are no longer limited by a specific space and time. On the contrary, political communication has been transformed into a continuum of information transfer that allows the coordination of individual agents on a global scale, as well as the flow of diverse experiences and knowledge.

In the same line, disintermediation affects "who is who" in political communication. The reduction of content production costs facilitated by Web 2.0 technologies and its own horizontal nature, makes citizens able to create content that is, in some cases, transmitted to a large number of people. This transformation of communication patterns has as its main effect, the circulation of content, messages, opinions, emotions and ideas that do not come from the mediating elites. From this angle, digital public opinion is more plural and/or has a greater capacity to introduce content for public debate.

After having travelled this road, we arrive at our research question mentioned in the introduction of this book: Does the context of disintermediation really generate a scenario closer to the ideal of public opinion? Our general interest revolves around, to what extent the development of 2.0 technologies and, consequently, the disintermediation of agents, messages and spaces, generate a more propitious scenario for the development of an inclusive and critical public opinion (top-right part of our boat). To ask ourselves about this is to reflect on how the rest of the actors (traditional actors) respond to the unforeseen consequences of technological development (upper right part of our boat). Unforeseen consequences that give citizens greater public prominence.

It is, in this sense, that the proposal of Habermas in "History and criticism of Public Opinion", help us as a beacon in the fog. In it, the author illustrates the contradictions that are expressed in the historical development of public opinion. These contradictions force us to be cautious imagining that the process of technological change we are experiencing will necessarily generate a more open communicative context. Habermas explains how, the emergence of bourgeois class was an advance on public opinion dominated by the social elites. That innovation is the critical and rational character of bourgeois public opinion. However, and as a necessary and inverse correlate, this public opinion responded to the interests of its own class and tried to legitim their values as general interest.

The other central element that we take from the work of Habermas to strengthen our analysis are the normative conditions that should form the basis of an ideal public opinion. Throughout this book, it has been pointed out how, in normative terms, public opinion should have a capacity to respond to problems or issues that arouse its interest, should be independent in the sense of not being co-opted and should be inclusive allowing the participation of a broad representation of citizenship.

Despite its positive aspects, agents' disintermediation is also characterized by a reaction of traditional agents (political parties, media, etc.) expressed in how and for what their representatives use social networks. As indicated in this book, traditional political parties in European countries such as Spain or Americans such as Colombia or Brazil use digital technologies as a mechanism to further disseminate their messages and proposals. It is the emergence of the so-called connective parities that opens a way of change since these are organizations open to debate with citizens and that allow their opinions to be included in their electoral programmes or be part of the political strategies of said parties. However, in practice, and after some years of existence of connective parties, we know that the representatives of these parties do not meet the criteria that defined their parties as connective (De Marco, Robles, & Gómez, 2019). Thus, for example, the political party Podemos (one of the recurrent study cases in this work) is characterized by establishing very little contact via social networks with their followers and by responding with a frequency as low as the rest of the traditional parties to the interpellations of citizens. Similarly, the various public debate campaigns initiated by this party in cities such as Madrid are transformed into spaces for debate with a very low level of effect on the public policies of that city.

This tendency leads us to consider that the "idea" of connectivity of the connective parties is nothing more than a strategy of absorbing the "ideal" of disintermediation. Behind an image of horizontality, traditional communicative strategies are hidden. These strategies consist of using social networks as a means of transmitting content developed by the party, added to a low interaction with citizens. That being the case, digital public opinion is dangerously moving towards a scenario very similar to that described by Habermas when speaking of bourgeois public opinion. That is, a scenario in which the interests of certain sectors are publicized as universal interests; the interests of all.

As we saw above, the privatization of public opinion is to pass on private interests (the interests of a group, political party or class) for the interests of all. The idea, although not the practice, of horizontality and inclusivity under which the connective parties are presented generates the false image of a public opinion in which the voice of all is represented. The feeling of being within a click of the public representative creates the expectation of accessibility and closeness when the reality is that these representatives are as far away as ever. With the idea of connectivity, traditional organizations give the appearance of universality to what is really the will of the representatives of these organizations. It is, therefore, a bourgeois public opinion (public opinion of class or status) clothed in digital clothing.

The disintermediation of messages also has a response from the agents (lower-right area of our scheme). Ideal public opinion, at least in the Habermasian version, does not only include the question of universality. Universality that, we have seen, is put into question by the transformation and strategies of digital games. Another key aspect is inclusion. Public opinion must be as inclusive as possible and give rise to the voice of all being considered. However, the disintermediation of messages and the individualized nature of this dimension of disintermediation generates a false image of inclusion and belonging that is very far from the ideal of public opinion.

As described in this book, disintermediation generates a very specific relationship with the messages and content that are the basis of public debate. Inclusion and belonging signify commitment and responsibility. Commitment to the causes that a citizen defends and responsibility with the objectives and results of the actions that are undertaken. However, unlike the activism that associated citizens with the causes that defended the classical social movements and the new social movements, the connective action generates an individual and egocentric link with the causes. Individual in the sense that the citizen who retweets a message or joins a cause does so with little connection with any organization or with other activists. That is, based on a relational commitment that can be null. Egocentric insofar as, not being committed to a common cause, the motivation for participation goes through personal motivations.

The last of the forms of disintermediation, the disintermediation of spaces is not free of shadows. As we have shown in this book, the disintermediation of time and space generates two key effects. The first of them is the acceleration of public debate times. Speed is, in many cases, the enemy of the reasoned elaboration of arguments. Retweet and "like" are not only a threat to public debate because they imply a weak commitment to causes and political initiatives. They are a threat because they are framed in processes too fast to be the result of rational judgement and elaborate argumentation. It is difficult to imagine a democratically innovative public opinion under the premise that the arguments and support that a cause gathers are made at the speed of a click. We consider it impossible to imagine that, under these conditions, the ideal principles of rationality and argumentation that we inherit from the bourgeois publicity described by Habermas can be fulfilled.

The disintermediation of spaces is accompanied, in many cases, by a tendency towards technological solutionism. That is, the belief that all problems have a unique and technical solution. It does not matter the specific characteristics of the subject or the event that we are dealing with. It does not matter if it is produced in a country in the Middle East or Central Africa. The moral, political and technological principles that will inspire the solutions offered by disintermediated public opinion will be univocal and, most of the parts, filtered by the sieve of Western culture.

However, the main threat to digital public opinion is polarization and, within it, the incivility. Polarization is characterized by the tendency to establish closed debates with those people who share opinions and preferences. Certainly, this is a tendency to some extent normal, as it is very likely that people with common attributes tend to establish stronger relationships (homophily). In terms of political communication, it is logical to expect that people in favor or against a certain opinion, tend to be separated. However, it is in contexts marked by extreme tendencies, that polarization is defined, not only by an intense communication with those who think the same, but by a distancing or disqualification from the other. This is a problem for political communication but it is only the tip of the iceberg.

A complete and ideal public opinion cannot renounce its critical dimension. That is, their ability to respond and oppose those processes and public decisions that are considered unfair. This critical dimension, as we know thanks to the work of Habermas, is structured, from the beginnings of modernity based on rational arguments and intelligible argumentation. It is precisely this central dimension of public opinion that is most affected by polarization and incivility. Polarization, as it has been described in this book, presents, as one of its central characteristics, the reinforcement of the starting positions of the people who are debating. Thus, while in our daily life in the off-line world, the interaction, not always intentional, with people with different opinions forces us to put together understandable and rational arguments, polarization generates, precisely, the opposite effect. In closed debates with people whose opinion is homogeneous, the tendency is to reinforce and maximize the shared arguments. Meanwhile, dissent or alternative perspectives are something eccentric. Thus, the higher the level of polarization, the lower the critical capacity of the public debate. This is due, from our point of view, to the fact that, the tendency in this type of debates does not reward creativity and rationality, but the arguments that reinforce and multiply what has already been accepted.

In this context, incivility generates the perfect storm by complementing the process you just described. If, as has been pointed out, polarization generates reinforcing arguments and restricts the critical capacity based on intelligibility and rationality, the lack of civility works as a closure to the reasons and opinions of others. As it has been explained in this book, the lack of civility does not consist only in insult or discredit, but in the exclusion of the other from the public debate. Thus, if polarization makes the critical and rational dimension of public debate weak, incivility raises walls that impede argumentation and debate.

The main empirical finding of our research is framed in this second scenario. We have identified, both in the Spanish and North American cases, as well as in other cases referred to in the literature, that the most active accounts in social networks are those that generate more polarization and less civic discourses. Meanwhile, less active accounts, foreseeably accounts associated with non-professional citizens or with insufficient time available to participate in political debates, are expressed more moderately and rationally. Therefore, when evaluating the impact of disintermediation on public opinion (top-right part of our boat) we must appeal to a more elaborate discourse than that of cyber-optimism or cyber-pessimism. We also know that the most common communication strategy in social networks like Twitter is to forward messages. That is, it is much more common to take a message already written (usually by opinion leaders or media leaders) and send it to friends and acquaintances. We also know that polarization is more common, if the incendiary messages come from social, media or politically respected figures. Thus, it is this type of very active accounts that are most interested in moving the practice of public debate away from the idea of critical and rational communication. Maybe, it's not a conscious decision. However, the empirical fact is that it is not the citizens, but their representative and mediating institutions that play against the ideal.

The corollary of our arguments, as well as the answer to our research question, is as follows. The reaction of traditional agents, maintaining traditional communication strategies and opting for polarization, together with the individualized and fleeting nature of the disintermediation of messages and spaces, generates a complex process of legitimation. Under the image of a horizontal and inclusive public opinion hides a process of perpetuation of power relations and reinforcement of the guidelines of the liberal political system; individualism and the banalization of politics.

We recall at this point Habermas's diagnosis about the bourgeois public opinion. According to our interpretation, the bourgeois publicity expresses a double contradiction. On the one hand, it is a commitment to a critical and argumentative public opinion. However, the first contradiction, under this nature was concealed an attempt to pass the class objectives by universal objectives. Public opinion is thus privatized. Likewise, the second contradiction, when public opinion is transformed into a mass opinion, ceases to be critical and begins to legitimize the power that the bourgeoisie is now part of.

The digitally measured public opinion seems to be handled in a similar way. Although, at first it was offered as an inclusive and critical space, the contradictions did not take long to appear. In the first place, the connective parties ceased to be inclusive and horizontalizing. Secondly, the communicative strategies of the main agents began to be polarized and to be incivil. Third, communicative practices were framed in fleeting communication processes, little elaborated and simplified (solutionism). All these contradictions, from our point of view, hide the potentialities that, like bourgeois criticism of power, could have at first the digital public opinion.

Far from wanting to be apocalyptic, we consider it our responsibility as social researchers to pay attention, not only to the positive effects of new forms of political communication, but also to the monsters that create the dreams of a technological future. This technological utopia often offers us a science fiction image in which Web 2.0 technologies will allow a citizen empowerment that will make it possible to fight against tyranny wherever it takes shape.

This cyber-optimistic utopia, although as we have seen, has good reasons to exist, coexists with a reality marked by insult, the adulteration of communication via fake news or with a polarization so negative and extreme as to make difficult the understanding between parties. The empirical research provided in this book, the result of the compilation of previous studies and some carried out by the research group to which the authors belong, uses Big Data and social networks technics to analyse the effects of disintermediation on real practices of the citizens (bottom right of our boat).

Thanks to our studies, we now know that the technologies allow public debate to defend the rights of racial minorities in the United States (#BringBackOurGirls), to give hope of democratic regeneration to Spanish citizens (#UnidosPodemos) or to fight for freedom of people in no democratic countries (#BlackLivesMatter). Disintermediation generates this. A kind of light of hope in times of public disaffection, a connective action that makes us think that we can change the world. However, and perhaps precisely because of this, it is especially dangerous. We refer to the fact that disintermediation generates the image of an open, inclusive and influential public opinion.

It is difficult to deny, if we assume the thesis of disintermediation, that social networks and other Web 2.0 technologies have generated a greater capacity for citizen response to issues that arouse the interest of public opinion. Likewise, the concept of disintermediation implies the assumption of an opening and inclusion of voices that, once, were hidden or, in the best of cases, represented by mass media or political parties. However, it seems that this culture of political communication 2.0, marked by disintermediation, is perceived more clearly by non-professional citizens of political communication than by professional politicians and/or great agents of communication. It seems as if we were witnessing the clash between two political cultures. The old bourgeois culture of mass communication and the new culture of digital communication that assumes as its own the idea of disintermediation. It seems as if the culture of mass communication were defended by traditional agents with tactics of confrontation and spectacle (see Chapter 2 of this book) and the new digital culture were supported by the citizens thanks the disintermediation of agents, messages and spaces (see Chapter 3). While mass communication agents try to maintain its status as central doers of communication, the new culture tries to open the way for new non-professional agents take part in an inclusive and critical communication system.

Because of all this, the upper-right part of our boat should be occupied by an image of the Greek goddess Athena.[1] Not only because in the current public opinion are diverse strategies in force, but because these strategies seem to be facing a battle for the prevalence of one or another culture. While, as noted in this book, in the field of economics this struggle seems lost, in political communication still resist some Spartans to the powerful onslaught of the army of Xerxes.

References

Benkler, Y. (2006). *The wealth of networks: How social production transforms markets and freedom*. New Haven, CT: Yale University Press.

Castells, M. (2009). *Comunicación y Poder*. Madrid, Spain: Alianza Editorial.

Coleman, J. S., & Coleman, J. S. (1994). *Foundations of social theory*. Cambridge: Harvard University Press.

De Marco, S., Robles, J. M. & Gómez, D. (2019). *Connective parties and communicative political compromises: The lack of real interaction*. Political Science Meeting, Brussels.

[1]Athena is, in classical Greek culture, the goddess of wisdom, but also of war and strategy.

INDEX

Printed by Printforce, the Netherlands